DFS

228

David Myhra

I0809088

Schiffer Military History
Atglen, PA

Opposite: Looking down on the general plan arrangement of the *DFS 228 V1*. The entire aircraft was wood constructed, and its simple monocoque fuselage was built in three sections. Digital image by *Mario Merino*.

Book Design by Ian Robertson.

Library of Congress Catalog Number: 00-106472

Printed in China.
ISBN: 0-7643-1203-0

We are interested in hearing from authors with book ideas on related topics.

Published by Schiffer Publishing Ltd.
4880 Lower Valley Road
Atglen, PA 19310
Phone: (610) 593-1777
FAX: (610) 593-2002
E-mail: Schifferbk@aol.com.
Visit our web site at: www.schifferbooks.com
Please write for a free catalog.
This book may be purchased from the publisher.
Please include $3.95 postage.
Try your bookstore first.

In Europe, Schiffer books are distributed by:
Bushwood Books
6 Marksbury Avenue
Kew Gardens
Surrey TW9 4JF
England
Phone: 44 (0) 20 8392-8585
FAX: 44 (0) 20 8392-9876
E-mail: Bushwd@aol.com.
Free postage in the UK. Europe: air mail at cost.
Try your bookstore first.

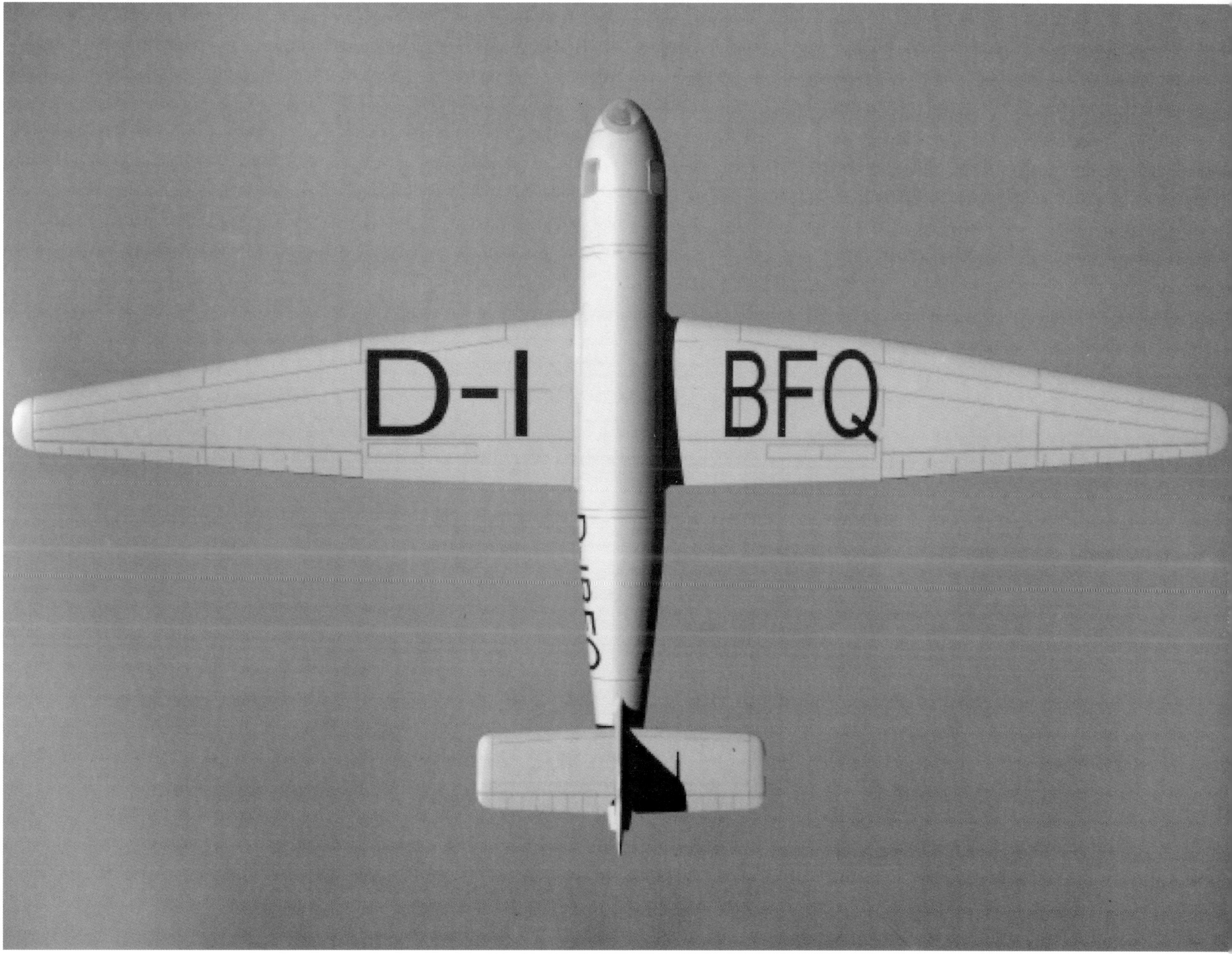
D-I
BFQ

DFS 228

Introduction

Beginning about mid 1940, engineers at the *Deutsches Forschungsinstitut für Segelflug*-Ainring (*DFS*) or German Gliding Research Institute-Ainring, proposed a single seat, bi-fuel liquid rocket powered photo reconnaissance aircraft for operating at ultra high altitudes. In essence, the *DFS* engineers were proposing a liquid rocket propelled sailplane which was to be flown in the prone position. In 1941, the *RLM* accepted *DFS'* proposal and authorized the construction of three prototypes to be designated as the *DFS 228*. But, as Germany's war time needs were expanding to the production of fighters and short range twin engine bombers, work on the *228* prototypes slowed down, and in fact stopped. Then, about 1943, the *RLM* authorized *DFS* to give their full attention to the liquid rocket- powered reconnaissance sailplane, and this is when *DFS* became actively engaged in the construction and testing of a bi-fuel liquid rocket propelled short range photo reconnaissance airplane which had sat on the shelf for three years. The *DFS 228* was intended to operate at heights up to 82,000 feet, or about 15 1/2 miles [25 kilometers] altitude. When operational, it would carry two state of the art *Zeiss* infrared cameras, allowing for a very wide area of ground to be photographed. No armament or armor was to be fitted to this flying machine. Eventually, the *RLM* ordered the construction of ten additional *228* aircraft, and these were known as the *DFS 228A*-0. A production line had been established at Griesheim, near Darmstadt.

But is this all there is to the *DFS 228?* Just another German wartime aircraft design powered by a single *HWK 509A* bi-fuel liq-

Felix Kracht's **first** ***DFS 228*** **prototype...the** ***V1*** **coded** ***D-IBFQ*** **as seen in level flight from its port side. The** ***V1*** **was completed in 1943 and non-powered flight testing was carried out at** ***Erprobungsstelle*****-Rechlin.** ***DFS*** **employed a** ***Do 217K V3*** **as the carrier aircraft. Digital image by** ***Mario Merino*****.**

The *DFS 228 V1* as seen in level flight but a bit above camera. The long dark band under its starboard wing root is its hydraulically articulated landing skid. The empty weight of the *V1* was 3,620 pounds and 9,280 pounds fully loaded. Digital image by *Mario Merino*.

uid rocket engine? Usually things are not what they seem, and this is certainly true for *Felix Kracht's DFS 228*. It was, by no means, just another bi-fuel liquid rocket propelled machine, or in this case a ultra high altitude photo reconnaissance machine. Let's go back for a moment and explore the real history, the impetus, motivating *Felix Kracht* and the reason for the *DFS 228*.

The May 1938 *Rhön/Wasserkuppe* Summer Sailplane Competitions And The Death of *Werner Blech* in a *Ho 3c*

Felix Kracht was a member of the 1930s generation of sailplane designers, builders, and flyers. This generation, born about 1915, missed the fighting of World War One but suffered its cruel aftermath in the form of the *Treaty of Versailles*, especially its ban on powered flight. *Kracht's* colleagues included the *Horten* brothers, *Walter* and *Reimar*, and a large group of individuals who would go on after their college days and become employed throughout the *RLM* and German aircraft factories. *Kracht* studied aeronautics in college and became a "*Flugbaumeister*," a coveted *Nazi Partei* sponsored program meaning "aircraft construction master." While in college he designed his own sailplanes, just as many other young men of his generation were doing, but he was better than most. These young men were all fierce competitors at sailplane competitions in the late 1930s. But another important item, they a were a tightly woven community of sailplane fliers who looked out for each other. It did not matter much the individual's politics: some were democrats, some neutral, and a few, like *Felix Kracht*, ardent

The *DFS V1* as seen from its nose port side and a bit beneath camera. In this first prototype the pilot sat upright, but seating would be changed to a prone position in later prototypes. Note the agricultural farm land beneath. The two round areas have been irrigated via a circular sprinkler. Digital image by *Mario Merino*.

The *DFS 228 V1*, flying above the German farmland, however, as it enters a cloud the farm land is obscured. *Felix Kracht* designed the *228* to give a flight range of 650 miles: 465 miles of intermittent powered flight and 185 miles of gliding flight. Digital image by *Mario Merino*.

Nazis. But what happened at the 1938 *Rhön/Wasserkuppe* was an event which would change *Felix Kracht's* idea of what real research was all about, *Nazism* or otherwise.

As the time for the August 1938 *Summer Sailplanes Competitions* at the *Rhön/Wasserkuppe* drew near, the *Horten* brothers were ready with two all-wing *Ho 3a* sailplanes: one for *Werner Blech* and one for *Heinz Scheidhauer*. In practice, they found that *Blech's Ho 3a* was slightly top heavy due to his heavier body weight, and as a result *Blech* was not doing well in tight turns. *Reimar Horten* felt that it was too late to make any major structural changes. But he did decide that the only thing to do to improve the *Blech's Ho 3a*'s performance and reduce its top heaviness was to give the machine a fixed pre-wing, or canard. *Reimar* changed the machine's designation, too, of *Blech's Ho 3a* to *Ho 3c*..."*C*" for canard. On the other hand, he was to fly in good thermals and during speed flights, *Blech* could remove the pre-wing, because top heaviness did not matter in those situations. However, when going for altitude, the pre-wing would make the *Ho 3c* more efficient, *Reimar*

Felix Kracht seen sitting in his sailplane design known as the "*Rheinland*." It was flown in a reclined or recombinant sitting position. His colleague, *Hermuth Ruhbrauk*, is standing to the right.

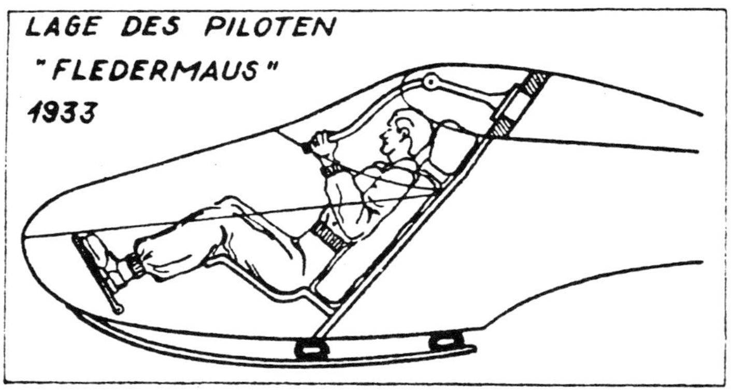

The *Fledermaus*, or field mouse sailplane of 1933, with its reclined or recombinant piloting position.

***Felix Kracht*, looking pretty salty when conversing with this author in late May 1986. Toulouse,France. Photo by the author.**

***Felix Kracht*, looking very dapper, about 1940.**

believed, and he wished that he had a pre-wing for *Scheidhauer*, too. But *Scheidhauer* did not want one, stating that its effectiveness had yet to be tested.

Both *Scheidhauer* and *Blech* did well during the first week of competition and placed well in the standings. The next phase of the competitions was altitude. But as the *Rhön's* second week began, the skies turned cloudy. Flying in cloudy weather did not bother *Blech* as much as it did *Scheidhauer*, and although both flew in it, it was *Blech* who was scoring points. Now, three days before the end of the 1938 *Summer Competitions*, *Blech* had accumulated seventy points to *Scheidhauer's* twenty-four. There was still time for *Scheidhauer* to catch up before the competitions were over. The weather was turning increasingly cloudy and then dark, as massive thunderclouds began to form.

As the time for the day's events drew near, *Blech* and *Scheidhauer* were next to be towed up into the sky. *Blech* told *Scheidhauer* that if he reached the blackening cloud formation first, *Scheidhauer* should not follow him and vice versa. *Blech* wanted to avoid any possibility of colliding with *Scheidhauer*, and he thought that since he had already scored so many points in reaching altitude, with one more try he just might win this phase of the *1938 Rhön/Wasserkuppe Summer Saiplane Competitions*.

At the start of the event, *Walter Horten* towed *Blech*, then *Scheidhauer*, into the sky, as usual. *Blech* entered the black clouds

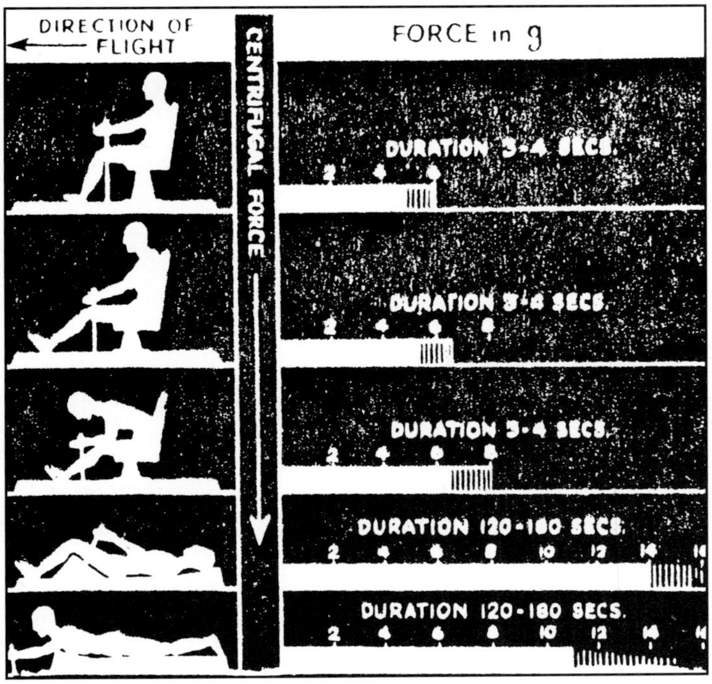

A pen and ink illustration featuring pilots in various piloting positions and the duration, in seconds, in the length of time they can remain conscious given high centrifugal force. *Dr.-Med. Siegfried Ruff's* research revealed that a pilot placed in an aircraft cockpit in the standard sitting position can withstand a high gravitational force for about 3 to 4 seconds while a pilot in the prone position, given the same high gravitational force, was able to withstand it for as long as 2 and 3 minutes.

***Dr.-Med. Siegfried Ruff*, right. Early 1940s.**

Fourteen year old *Reimar Horten*, seated in a training sailplane, and his friend *Siegfried Ruff*. *Rhön/Wasserkuppe*, 1929.

***Dipl.-Ing. Eberhard Meyer,* assistant to *Felix Kracht*. *DFS-*Ainring. Summer 1940.**

first, and a few moments later disappeared. *Scheidhauer* was several seconds behind him, and he too entered the black clouds and disappeared. Other sailplanes also entered the Storm clouds (*Walter Horten* told this author in the late 1980s that he had counted 18), but many flew right out again. This cloud, they had quickly discovered, was filled with lighting. *Kurt Hieckman*, a 3rd member of the *Horten* sailplane team, was piloting a *Ho 2* sailplane, and though experienced in blind flying like *Blech* and *Scheidhauer*, found himself rising instantly to 16,400 feet and decided immediately to leave the thunder clouds. Inside the cloud formation it was all black, illuminated only by repeated flashes of lighting. His *Ho 2* was rising at a rate of more than 2,000 feet per minute. *Hieckman* reported that more than once his instruments registered a rate of climb of 4,000 feet per minute.

On the ground *Walter* and *Reimar Horten* watched the storm clouds nervously. Suddenly, an *Ho 3* emerged and then floated toward the ground, making a great circle around the *Wasserkuppe*. As it came closer, *Reimar* saw that the cockpit canopy cover was missing, and that something was fluttering in the wind behind the sailplane. It was *Blech's* parachute, ripcord, and sack. Slowly the *Ho 3c* floated down, until it finally crashed flat on the open ground. Although the parachute had been released, everyone thought that *Blech* was still wrapped up inside it. Running over to inspect the machine, *Reimar* and the others discovered that it was not *Blech* hanging from his sailplane. Nor was he even in the cockpit. Checking the *Ho 3c's* barometer, *Reimar* discovered to his horror that the needle had gone beyond 26,250 feet, the last point on the instrument's dial. Apparently, *Blech* and his *Ho 3c* had gone even higher. Moments later, *Sepp Kunz* of the *NSFK*, a long time friend of the *Hortens*, told them that *Blech's* ice coated frozen body had come out of the sky, landing on a stony outcropping of the *Wasserkuppe*.

***Dr.-Med. Siegfried Ruff*, right, and *Professor Dr.-Med. Klein*. Late 1987.**

A revisit to *Eberhard Meyer*, at his home in Heiligenberg, West Germany, one year later. May 1988. Photo by the author.

***Eberhard Meyer*, at his home in Heiligenberg, West Germany, May 1987. Photo by the author.**

Looking up to the skies again, the crowd now saw a parachute descending with an apparently lifeless body dangling beneath. It was *Scheidhauer*. A few moments later *Scheidhauer's Ho 3a* was seen floating down. It hit the ground among the *Wasserkuppe's* pine tress with a crash as it broke apart. Rushing over to *Scheidhauer*, they found the man still alive but unconscious. He and his parachute were coated with ice. *Reimar* and *Walter* released *Scheidhauer* from his parachute harness and rushed him and *Blech's* lifeless body to the hospital in the nearby town of Tann. *Scheidhauer* was found to have severe frostbite over his body, and would have to remain hospitalized for more than six months. Two fingers of his right hand were so badly frozen that they had to be amputated.

Later, when *Scheidhauer* was well enough to talk, he described his ordeal in the black clouds. He and *Blech*, he believed, may have reached an altitude of 32,800 to 36,100 feet. Inside these clouds there was total blackness, and the temperature was well below freezing. It began to hail, and the stones were as big as golf balls. Moments later his cockpit canopy was shattered, as well as the plywood covering over the leading edge of the wing. He knew, *Scheidhauer* said, that he was at a great height, but he did not know quite what to do about it. He was wearing a seat belt and a parachute that would release automatically after he jumped out of his *Ho 3a*, so he decided to jump. Preparing to leave his *Ho 3a*, which by now had been perforated by hailstones, he grabbed his barograph so he would at least get some points for the ultra high altitude he had reached. As soon as he jumped, he apparently passed out. A later inspection revealed holes the size of a man's fist throughout the entire upper surface of the all-wing sailplane.

***Felix Kracht* in late May 1986. Toulouse,France. Photo by the author.**

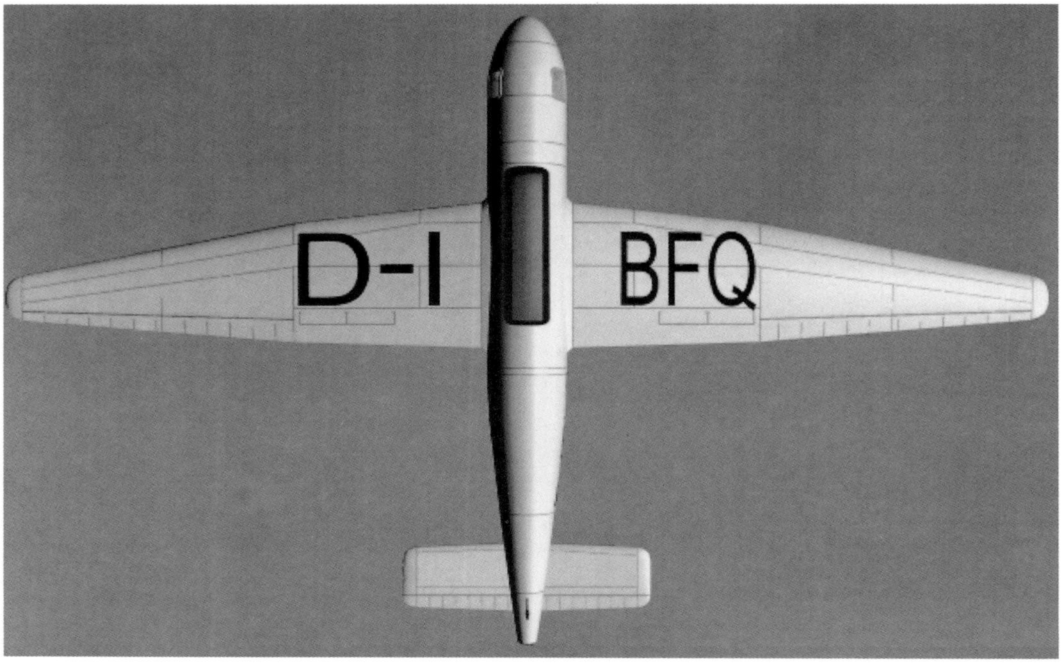

The bottom side of the *DFS 228 V1*. The rectangular box shown between the wings is the *V1's* hydraulically articulated metal landing skid. The only item housed in the tail section was the *HWK 509D's* combustion chamber. The pumping section was located mid-fuselage. Digital image by *Mario Merino*.

They now knew what had happened to *Scheidhauer*, but the events surrounding *Werner Blech's* death were not so clear. It appeared that he had stayed with his *Ho 3c* up to and beyond 26,250 feet in what was called then "static" standing waves. Some eighteen other sailplanes had been in the thundercloud about the same time, though not near the same altitude. Three other sailplane pilots had bailed out, but landed with no harm or difficulties. Apparently, *Blech* had left his *Ho 3c* at great altitude. In the darkness of the clouds, he may have been struck by the wing of his own *Ho 3c*, or by the wing of another sailplane as he was on his way down by parachute. He had landed with his parachute intact. Later, at the hospital in Tann, the *Hortens* were told that *Blech's* neck and nose had been broken.

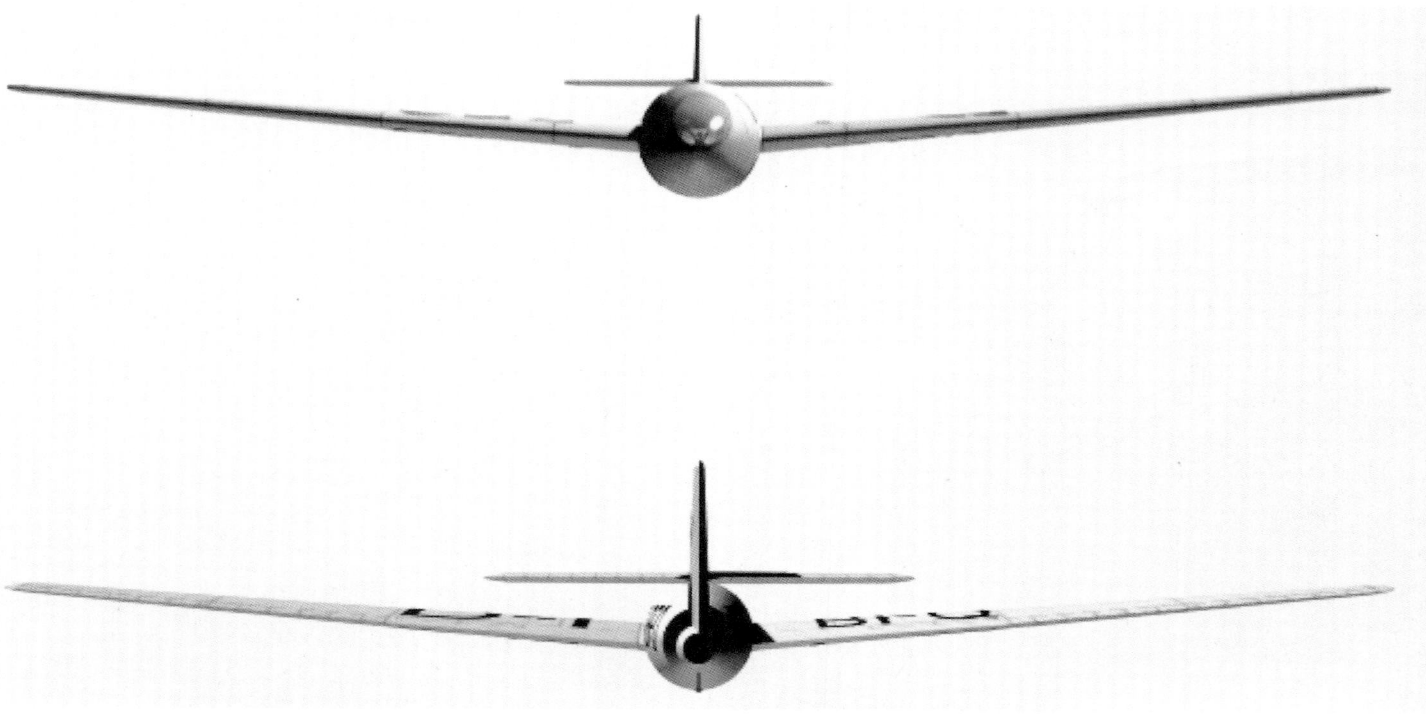

A view of the *DFS 228 V1*...directly in front and from the rear. The 57 foot, 1/2 inch long wing span contained a single laminated wooden spar extending wing tip to wing tip. It also contained wooden ribs and full plywood covering. Digital image by *Mario Merino*.

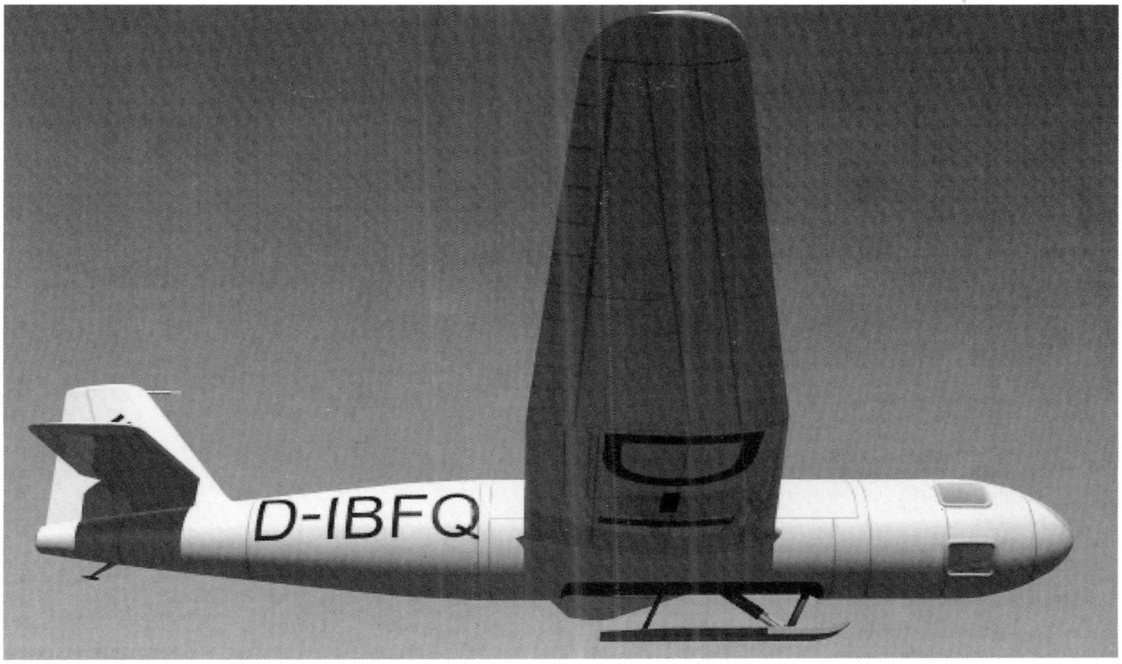

The *DFS 228 V1*, its starboard side, as seen from a slightly below position. The machine is seen with its hydraulically articulated metal landing skid fully extended. Digital image by *Mario Merino*.

Months following the *1938 Rhön/Wasserkuppe Summer Competitions*, *Felix Kracht* began drafting up plans for a detachable pressurized cockpit cabin for his fellow sailplaners, and he never gave up this idea for protecting the lives of ultra high-flying sailplaners. As the design progressed at *DFS*, the *RLM* gave the radically new sailplane the designation number *DFS 108-054*. This was in 1939. *Kracht* had convinced the *RLM* that substantial progress in the field of glider soaring had been made with the discovery of the so-called "static" standing wave, which reached considerable heights for their areas. The heights that sailplane gliders were reaching were increasing each year. In Autumn 1940, 38,058 feet [11.6 kilometers] was the height limit which could be reached with the use of an ordinary oxygen mask. However, the fields of "static waves" were considered to reach considerably higher, therefore, *Kracht* argued, the construction of a glider with a pressure cabin was imperative. *Kracht* found difficulties in designing such an ultra high altitude sailplane for the lowest possible sinking speeds, providing the highest aerodynamic quality, cockpit ventilation, and with a low wing loading. *Kracht* and his able assistant *Eberhard Meyer* had to deal with the issue of the cockpit cabin's weight. They found that a closed cabin with a circular ventilation system could be realized, and their tests showed that such an air tight cabin could be constructed from wood. So, with the *RLM's* approval for prototypes designated *DFS 54*, work started in 1940. *Kracht* said that design and development

A full port side view of the *DFS 228 V1*. Notice that its landing skid slightly extends out beyond the wing's leading edge. The wooden center fuselage section contained 2x*Zeiss* infra-red cameras, tanks for the *C-Stoff* and *T-Stoff*, and the *HWK 509D* bi-fuel liquid rocket motor. The *509D's* combustion chamber was located in the tail section, connected via long extension pipes to the pumping section located mid-fuselage. Digital image by *Mario Merino*.

A view of the prototype *DFS 228 V1* ultra high photo reconnaissance aircraft wing tip to wing tip out in front and a little beneath the camera. *Felix Kracht* designed this machine to reach altitudes of 75,000 to perhaps as high as 100,000 feet. Digital image by *Mario Merino*.

work were completed, and construction of the prototype suffered constant interruptions. It was about 80% complete when the project was put on the shelf and finally abandoned because of the widening conflict spreading through Western Europe.

The *DFS 54* survived the war, and all completed sections of the *DFS 54* were located at *DFS*-Prien. *Kracht* estimated that completion of the prototype would take about 2 to 3 months work. He told his American interrogators that the *DFS 54's* technology was out of date; nevertheless, a pressurized cabin should be constructed for a prone position of the pilot based on their experience with the *DFS 228/346. Kracht* argued that if the Americans would allow him and his workers the time and expense then they could have the pressurized sailplane cockpit ready for flight in about six months. Completing the sailplane in its present condition would require approximately 3,000 man hours. However, if bringing the *DFS 54's* pressure cockpit up to the state of the art, then it would require approximately 800 hours of design work and 4,000 to 4,500 construction man hours. The *DFS 54* prototype was not assembled or re-engineered, but scrapped and burned in the base wide clean up.

The *DFS 228 V1's* underside showing how its hydraulically articulated metal landing skid fit flush and tight within the fuselage's center section when fully retracted. Digital image by *Mario Merino*.

***DFS 228* Prototype - Dimensions and Weights**
• Wing span - 57 1/2 feet
• Wing area - 323 square feet
• Wing aspect ratio - 10
• Length, overall - 34 feet 7 inches
• Weight, structure - 2,980 pounds
• Weight, maximum flying - 9,280 pounds
• Weight, landing - 3,760 pounds
• Speed, maximum - 565 miles per hour
• Speed, maximum cruising - 560 mph
• Fuel - approximately 684 gallons
• Range, maximum - 650 miles (465 miles of powered flight plus 185 miles of gliding flight)
• Service ceiling, absolute - 83,000 feet
• Average operating altitude - 75,000 feet
• Power Plant - 1x*HWK 509D* bi-fuel liquid rocket engine of 3,300 pounds thrust at sea level and 3,630 pounds thrust at 40,000 feet, and fitted with a compressed air system pump for pressurizing *C-Stoff* and *T-Stoff* fuel feed lines at ultra high altitudes.
• Launch altitude - 33,000 feet
• Launching aircraft - *Do 217K V3*

Method of Operation

To economize fuel and to better attain a higher operational range, the *DFS 228* would be carried on the back of a *Dornier Do 217 K3* up to 33,000 feet altitude. At this point the *DFS 228* would be released for its operational flight, and the lighting of its 1x*HWK 509D* bi-fuel liquid rocket engine would take effect. Within minutes the *228's* service ceiling of 75,000 feet would be reached, and from then on it could freely operate within the "safety altitude" of 39,400 feet. Anything beyond 39,400 feet was outside the limits of Allied fighters. Approximately, a range of 465 miles was expected to be obtained by a process of steep climbs and shallow glides. An additional 185 miles could be obtained in the *228's* final glide down from its "safety altitude" of 39,400 feet. Thus, the extreme range of the *DFS 228* from the moment of its release could be as much as 650 miles.

Fuselage

The *DFS 228 V1's* fuselage was constructed in three sections, each of which was symmetrical about its longitudinal axis. The forward section, a self contained unit forming the pilot's cockpit cabin, was of ellipsoidal shape. The center section extended from the cockpit to the trailing edge of the wing. It was cylindrical in shape. The section aft the center section narrowed down to the end of the fuselage.

Cockpit

The cockpit was designed as a pressure chamber to maintain atmospheric conditions equal to 26,000 feet altitude. The inner shell was insulated against cold temperatures by four layers of aluminum foil sandwiched between it and the outer wall, with the total gap/space between the two shells being two inches. The cockpit was connected to the center fuselage with four explosive bolts. Controls and electrical connections were grouped together to facilitate assembly, as well as to make for a quick release.

The nose of the cockpit was formed by three molded plexiglass panels mounted in a cast *Silumin* ring. There was a space between the panels for the circulation of hot air. The *Silumin* ring was supported by the structure of the pilot's couch/platform, which was fastened to the rear bulkhead. The platform itself was built on a tube, which acted as a compressed air cylinder. Auxiliary equipment, such as air conditioning and heating apparatus, radio, and batteries, was carried on the platform framework. Access to the cockpit was gained by sliding the nose forward about 3 feet.

The *DFS 228 V1* seen coming on and featuring its port side. Because of its light weight (3,760 pounds without its *HWK 509D* rocket fuel of *C-Stoff* and *T-Stoff*), the *228* could be expected to maintain its altitude by relying on the up drafts from thermals just as high-performance sailplanes do. Digital image by *Mario Merino*.

Special care was taken in the design of the pressure cabin, which had, of necessity, to be simple, cheap to construct, and lightweight. As a result, it was reported that there were a number of pilot discomforts due to the consideration of keeping costs as low as possible. Nevertheless, discomforts were not sufficient to impair the pilot's efficiency.

The pilot of the *DFS 228* breathed pure oxygen, and had an ordinary mask, slightly modified by the fitting of a "*Silicagel*" capsule at the outlet to absorb exhaled moisture. Cabin air leaks were to be compensated by exhaled air, supplemented by a battery of compressed air bottles. Temperature and humidity were to be kept constant by a small air conditioning (AC) unit on board. This AC unit consisted of a ventilator driven off the *DFS 228 V1's* electrical circuit, which was to circulate the cold damp air through a drying capsule, heater, and finally delivering the warm air via the plexiglass nose cone. The heating element was an aluminum tube in which "short burning" cartridges were ignited. The heat generated would have been transmitted by longitudinal fins to the air which was induced to flow over the AC unit.

Of major interest was how the *DFS 228 V1* handled emergencies when the pilot would have to immediately leave the machine at high altitudes. The *228* had a pilot ejecting device for use in emergencies due either to fire in its *HWK 509D* bi-fuel liquid rocket engine or a sudden drop in cabin pressurization. *DFS*-Ainring's designers believed that the quickest way to return the aircraft's pilot to a "normal altitude" was for him to jettison the entire cockpit cabin. In order to immediately jettison the cockpit, the *228 V1* was fitted with four explosive jettisoning bolts designed by *Karl Bütter* of *Heinkel AG*. In addition, the whole cockpit cabin would be automatically jettisoned when sensors sensed a sudden drop in cabin pressurization. This feature was designed into the *228 V1's* cabin in order to protect an unconscious pilot.

The *DFS 228 V1* seen going away and featuring its starboard side. Powered flight had not been attempted by war's end on 8 May 1945, due to continuing failure in maintaining pressurization in its wooden cockpit cabin. Digital image by *Mario Merino*.

The starboard side of the all wood construction *DFS 228 V1*. In this early version the pilot flew the aircraft sitting in the standard upright position. Only with the *V2* and *V3*, was the pilot to be laying on a couch and flying the aircraft in the prone position. Digital image by *Mario Merino*.

The clean, uncluttered underside of *Felix Kracht's* ultra high altitude *DFS 228 V1* photo reconnaissance aircraft.

When the four jettisoning bolts were fired the cockpit cabin fell clear of the fuselage. As this was happening, a small stabilizing parachute was released from its parachute box at the outside rear of the cockpit cabin. As soon as external atmospheric pressure exceeded the internal pressure of the cockpit cabin, the seal around the plexiglass nose cone was opened. As the falling cockpit cabin reached a pre-set altitude, the pilot's couch would be pushed forward and out of the cockpit automatically by compressed air. As this was happening, the pilot's safety harness/belts and oxygen supply would be disconnected, and his parachute automatically opened, too.

Prone Position Piloting

Although the initial design for the *DFS 228 V1* required that it be flown in a conventional seating arrangement, *DFS* engineers all along had in mind that the pilot fly the aircraft in a prone position laying on a couch. The purpose was to make the diameter of the *228 V1's* fuselage as narrow as possible to improve performance. In addition, a small circumstance would make it more easy to seal the cockpit cabin. *DFS'* engineers had considerable experience with prone position piloting, several engineers having built sailplanes with this piloting position while still in college. It is estimated that *DFS* had made about 40 test flights with their *228VI* by war's end.

From this view it is easy to see that the high altitude *DFS 228*, with its long wing span and high aspect ratio of 10.0 was really a rocket-powered sailplane.

A view of the *DFS 228 V1's* tail section. The tail section contained only the *HWK 509D's* combustion chamber and tail skid. Pipes and tubes extended back to the bi-fuel liquid rocket motor housed in the center section of the fuselage.

Center Fuselage Section

Originally this section was of a twin wall wooden monocoque construction. It was modified because of the interruptions in structure needed for its retractable landing skid, the *HWK 509D* bi-fuel liquid rocket engine, and the *228's* one piece wing spar. Those parts of the *HWK 509D* in the center section included the main engine unit, such as steam generator, turbine, pumps, control valves, pressurizing fuel feed lines at ultra high altitudes, and electrics. Forward the main engine in the center section were the bi-fuel liquid tanks.

The *DFS 228 V1* landed on a retractable metal skid. It was attached to two fuselage longitudinal members on the under side of the fuselage shell.

Access to the *HWK 509D* bi-fuel liquid rocket engine was obtained from the upper part of the fuselage. A large, one piece, stressed, removable panel on the upper part of the fuselage was lifted off the fuselage walls by quick release clips. In addition, tanks for *C-Stoff* and *T-Stoff* and the twin *Zeiss* aerial photographic cameras were all located in the center section, as well.

Rear Fuselage Section

The rear fuselage section was a simple, wooden semi-monocoque shell carrying the tail assembly, the *HWK* rocket's combustion chamber, and the retractable metal tail skid. Since the combustion chamber was far removed, the rocket unit's main engine, the combustion chamber, without its thrust tube, was secured to a bulkhead for support.

Wing

The *DFS 228 V1* employed a cantilever wing which was built around a one piece wooden spar running wing tip to wing tip. For ease of

The *DFS 228 V1's* cockpit cabin had a single round plexiglass window in its nose. Two small, rectangular plexiglass windows were installed...one each on the port and starboard sides.

The nose starboard side of *Felix Kracht's DFS 228 V1*. Its pressurized cockpit cabin was designed to maintain a pressurization equal to an altitude of 26,000 feet. *Kracht*, *Meyer*, and *Mödel* had great difficulty in keeping the cabin pressurized due to the failure of the seals and the fact that the cockpit had to be constructed out of wood.

construction and economy of material, the spar flanges were of constant thickness, and the spar itself was built up from six double "T" beams. When finished, the spar was unusually wide, strong, and capable of high torsional loads. It was also very rigid, thus rejecting bending. The fabric covered ailerons were of a high aspect ratio (long and narrow), and extended over more than one half of the wing's trailing edge in two sections. The inner aileron served as a landing flap during landing. Lift spoilers were also fitted to both upper and lower wing surfaces. Slatted air brakes were fitted to control the *228 V1*'s gliding speed.

Fuel System

Normally, high-altitude flight imposed considerable difficulties in feeding liquid fuel because a normal suction pump was incapable of reducing the pressure at the delivery pipe below the external atmosphere, and therefore, vent pressure. *DFS* engineers considered using a pitot-head pressure, but later found that this form of pressure was inadequate at ultra high altitudes. Instead, the engineers designed a system for fuel delivery which used compressed air in assisting the fuel feed flow.

***Professor Walter Georgii* (1888-1968) in the early 1960s.**

***Professor Walter Georgii* (left) with *Hanna Reitsch* (middle).**

Professor Walter Georgii **(left) on the** ***Rhön/Wasserkuppe*** **pre-war with** ***Hanna Reitsch*** **and** ***Wolf Hirth*** **(right).**

RLM's **director of Research....*Adolf Baeumker*** **(left) and the** ***RLM's*** **Technical Director...*Ernst Udet*** **(right).**

Power Plant

The *DFS 228 V1* was powered by a single *HWK 509D* bi-fuel liquid rocket engine similar to the unit used in the *Me 163B* interceptor. The *HWK* unit was unaffected by altitude and/or speed when fitted with a fuel line pressurizing system, unlike a piston engine, and *DFS* engineers found this characteristic to be desirable.

State of Development

At war's end in early May 1945, *DFS*-Ainring had built three *DFS 228* prototypes. The fourth prototype was to have been fitted with their latest cockpit cabin design...the one with a full plexiglass nose cover as seen on the *DFS 346*. In addition, this cockpit cabin employed all the safety features for ultra high altitude flight, including the explosive bolt release of the cabin from the center fuselage. *DFS* engineers were also working to improve the *228's* maximum forward speed and refining of the aircraft's design to give a flatter gliding angle.

Two once powerful men within the *RLM*: *Roluft Lucht* (left) and *Gottfried Riedenbach*. April 1941.

Number of *DFS 228s* Surviving Post War

One complete example of the *DFS 228* is known to have been obtained by *General George McDonald's* U.S. Army Technical Intelligence people at war's end. It was the *DFS 228 V1* coded *D-IBFQ*, and was found at *DFS*-Ainring in the U.S. Zone of Occupation. When it was found the machine was undamaged, and it was disassembled and removed by U.S. Army trucks on 18 June 1945 to the U.S. Air Technical Intelligence Unit, Stuttgart. Then, as of 14 June 1946, the *228 V1* was in Farnborough, England, for further investigation and study. The report of this British investigation is known as *EA 280/1*. The agreement between England and the United States regarding the transfer of the *228 V1* to Farnborough is unclear. In March 1947, the *V1* was photographed in a scrap heap thought to have been located at Farnborough. But then, there are reports that this specimen was to have been transferred to *Slingsby Sailplanes Limited*, Kirkbymoorside, Yorkshire, about this time, too. *Slingsby's* engineers are known to have submitted a design for a "stratospheric" research sailplane in response to a request for proposals from the U.S. Air Research & Development Command. This request for proposals (RFP) was *EOARDC-54-41*, and *Slingsby's* response was known as the *Slingsby T-44*. The *T-44* was a two seater with side-by-side seating and a tricycle undercarriage. It is also unclear whether or not the *DFS 228 V1* was to have been shipped to the United States after Farnborough and *Slingsby* had finished with their investigations. Nevertheless, the *DFS 228 V1* appears to have been

***General Ernst Udet* visiting *DFS*-Darmstadt in Summer1940. *Udet* appears wearing the white stripe of a senior officer down his pants leg. Behind *Udet* is *Professor Georgii*. The tall person behind *Udet* is *Hans Jacobs*. To the far left in the photo, the second man is *Eberhard Meyer...Felix Kracht's* assistant.**

scrapped in Great Britain some time in 1947. Photos, taken some time in 1947, show the *DFS 228 V1* down on the ground, its wings detached and laying nearby. This scene's location is unidentified, but is somewhere in Great Britain. Upon closer inspection, the *228 V1*'s fuselage aft its *HWK* engine head and bi-fuel liquid rocket fuel tanks appears to have been broken, and its entire vertical stabilizer/rudder is missing. Overall, the remaining aircraft shows gross neglect in previous shipping and handling and a general deterioration. This high flying bird would never experience that thrill of flying again, and it would have taken a considerable amount of restoration just to make it presentable as an indoor static display.

General McDonald reported that his Technical Intelligence field team also recovered the detachable pressurized cockpit cabin intended for the *DFS 228 V3*. This cabin has never been seen again, and it is unknown if it even left the Stuttgart location of the U.S. Army Technical Intelligence facility.

Summer of 1940. *Ernst Udet* is shown in the left center of the photo facing the camera. Behind him and to the left is *Professor Georgii*, identified by his head of white hair. Also behind *Udet* and to the right is *Hans Jacobs*, the tall one. Next to the right of *Jacobs*, wearing the hat with hand in trouser pocket, is *Felix Kracht*.

***General Ernst Udet* casts a suspicious eye toward *General Erhard Milch*. Neither trusted the other. *Udet* would commit suicide in November 1941, and *Milch* would take over his Technical Department. *Milch*, post war, was found guilty at Nuremburg and spent twenty years behind bars at Spandau-Berlin. *Udet* felt a kinship with the staff of *DFS*, while *Milch* viewed the organization as just another of the *Reich's* research institutes.**

In Their Own Words...Interviews with Individuals Connected with the *DFS 228* Project

In the mid 1980s, this author met and spoke with several former members of the *DFS 228* design and construction team and DFS in general. Eight of these interviews are included here, and include:

- *Felix Kracht - DFS 228/346* Chief designer and project director;
- *Eberhard Meyer* - Former head assistant to *Felix Kracht;*
- *Hans Jacobs* - Head assistant to *Alexander Lippisch* and designer of the *DFS 230* transport glider;
- *Rolf Mödel* - Constructor of the several *Kracht* designed detachable pressurized cockpit cabins;
- *Werner Nestler - DFS'* specialist on the *HWK 509D* bi-fuel liquid rocket-propelled engines;
- *Hans Niess* - Assistant to *Felix Kracht*, and project leader on the *Me 328* pulse-jet miniature fighter interceptor;
- *Eric Klöckner - DFS* test pilot, and one of the test pilots in the *DFS 228 V1* program;
- *Hermann Zittter - DFS* test pilot.

Felix KRACHT
21-23 May 1986
Toulouse, France

MYHRA - Very little is written on the history of the DFS. *Professor Lippisch, Jacobs, Georgii,* and so on have said very little about

An Allied aerial photograph of *DFS*-Ainring, its facilities and grass runway (seen in the center of the photograph), taken 9 April 1945, or about a month before Germany's unconditional surrender on 8 May 1945.

their work at *DFS*. Former *DFS* test pilot *Karl Schieferstein* is probably the only exception.

KRACHT - *DFS* was an organization, a society, which had been created around 1925, with its laboratories in Darmstadt. They were interested initially in soaring and meteorology. In meteorology, *DFS* probably led the rest of the world, because they detected not only that there was a horizontal flow, but also a vertical flow of air masses.

MYHRA - Was not *Dr. Georgii* a meteorologist, and because of his interest in weather, he also was available to assist sailplaners, and so on, regarding weather conditions?

KRACHT - Yes. After *WW1* any activity in motor-flying was prohibited by the *Treaty of Versailles*. So it was quite natural that anyone who wished to fly with an engine now had to fly without. About this time—I think that it was about 1920-1921—at the *Rhön*, they started the first soaring activities. At this time *Georgii* was at the *Rhön* as an advisor. So, too, were the great sailplane pioneers, like *Klemperer*, *Martens*, and others. There were a lot of good people translating their aeronautical ideas into sailplanes.

MYHRA - Not yet *Lippisch*?

KRACHT - No, *Lippisch* was not yet there. Later on, though. I do remember *Robert Kronfeld*. But at this time the sailplanes which really made long distance flights were those built and flown by *Wolfgang Klemperer* from *Gothaer*-Gotha. Soaring developed more and more. Student groups from the different technical universities, particularly from the Darmstadt area, became very active. From this was created a society which was called the *Deutsches Forschungsinstitut für Segelflug* (*DFS*), or German Gliding Research Institute. So *DFS'* background comes from Germany's post *WW1* interest in soaring and their need to know more about and better understand meteorology. Then, during the 1930s, especially after 1933, *DFS* was charged with developing even bigger sailplanes. For example, this included the *DFS 230*, a transport designed by *Hans Jacobs* which could carry 10 to 12 fully equipped soldiers. From then on *DFS* was heavily involved in the development of transport gliders, soaring, and related research. This is about the time I joined *DFS*. The problem, however, was how to tow such large transport gliders in bad weather and darkness. At this time the ceiling of most glider aircraft was around 7,000 to 8,000 feet. During bad weather, which is frequently at the same height, these trans-

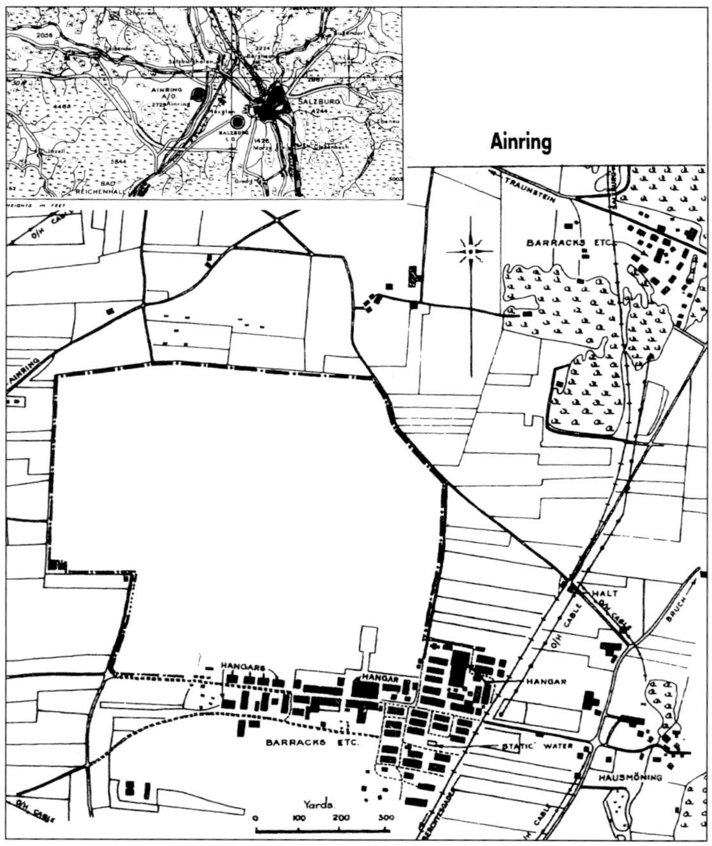

A pen and ink close-up illustration of *DFS*-Ainring's facilities, hangars, barracks, grass runway, and so on by the American military. The insert at top left shows the relationship of Ainring to its neighboring towns, such as Salzburg and Bad Reichenhall.

At the end of this road was *DFS*-Ainring. It was located in a beautiful mountainous area of Germany. *DFS* sort of planned it this way when they moved to Ainring from Darmstadt. War weary *RLM* senior officers from Berlin could come to the peaceful Ainring area and relax. Afterward, in thanks, they'd be more inclined to increase *DFS'* research and development funding. It worked, too. Photo by author.

One of the many mountain scenes around *DFS*-Ainring. This photo was taken on a rainy day by the author.

port gliders were right in the middle of the bad weather. *DFS* was given the task of trying to solve the problem of towing transport gliders in all weather conditions. Thus, we had come to the conclusion which we called the "coupling possibility" in flight could work. That is towing one aircraft behind the other with a rigid towing apparatus. This had been tried, but not without some difficulties.

MYHRA - Was this coupling also called "rigid towing?"

KRACHT - Yes. In Germany we usually referred to it as "rigid" towing. In addition, we experimented with combined flight controls and combined landings...the "*Mistel*" program. And then came the next question, is it possible to fuel aircraft under such conditions, in flight? First, this means overall, that you will not start together, but the aircraft needing fuel would have to catch the tanker in the air and then go over to rigid towing. When I joined *DFS*, refueling was my first assignment. I was told to develop a method so that one aircraft could refuel the other while in flight. My superiors thought that the aircraft requiring fuel should couple to the tanker via rigid towing. Many other ideas then derived from this task. For example, I was able to obtain many patents. To see how refueling is done today, well, it's still so damned complicated. If it was done on a rigid basis it would be much better and safer. For example, neither the front aircraft nor the rear aircraft would have to watch everything while refueling as it is done today. It would be a completely "hands off" arrangement. Back at *DFS*, from that time on in the late 1930s, we had come forward with many more ideas on towing. For example, the "father and son," or "*Mistel*" was de-

A very poor quality photo of the scenic snow-covered mountains surrounding *DFS*-Ainring.

A poor quality photo of the accommodations available for visitors at *DFS*-Ainring. With this powerful view it is easy to see why senior officials from the *RLM*-Berlin found reasons to go and visit *DFS*-Ainring.

Visitor's accommodations at *DFS*-Ainring in the winter time.

A very poor quality summer time view of the visitor's accommodations at *DFS*-Ainring. Notice the large in ground swimming pool.

veloped. This included the *Ju 88* as a separate aircraft, while the other aircraft was atop the *Ju 88*. As this program evolved, I remember, one particular flight which I tested this combination, flying only from a *Bf 109*. There was nobody in the *Ju 88* beneath me. I had just taken off from *DFS*-Hörsching, and I received a call over my radio that a great bomber pack was coming over, north of where I was flying and I was ordered to land immediately. So I landed, and during the landing one of the tires on the *Ju 88* blew out. Once I stopped the *Ju 88* would not roll again due to the blown tire, so my people came to me in a motorcycle with side-car. I climbed out of the *Bf 109* and into the side-car, and we sped back to the aerodrome. Then there occurred a curious phenomenon. This airport had two parallel runways, and then came those *B-17* bombers with their *P-51 "Mustang"* fighter escorts. Well, the *"Mustangs"* came by to look at what was on the runway. They flew up and away, but turned around and then they came back, shooting their cannons. At that time at DFS-Hörsching there had been installed at both ends of the runway some 20 mm anti-aircraft cannons on turrets with four barrels each. One *"Mustang"* was shot down near one end of the runway, while another *"Mustang"* was shot down at the opposite end. From that day on, we were never bothered again by *P-51s*, nor by those *B-17* bombers. So this day was the first and last time we had been attacked.

MYHRA - Tell me, how did the idea for the bi-fuel, liquid rocket-powered *DFS 228* come about?

KRACHT - There were people in a department, which was called LC-*1* in the German Air Ministry (*RLM*), and discussions and conversations about what could be done and what could be interesting to do in terms of aeronautical research. This continued to go on even though our war was lost once America entered the conflict in

A supper-time gathering of *DFS*-Ainring senior officials and staff in 1941. Left to right: *Professor Georgii, Pohl, Niess, Geul, Nestler, Mödel, Peter, Kick*, and *Fraulein Bohm* (*Felix Kracht's* ex-wife)

The same *DFS*-Ainring group but in different seating positions. Left to right: *Niess, Geul, Nestler, Mödel, Professor Georgii, Pohl, Fraulein Bohm, Peter, Kick, Wurzbach*, and *Meyer*.

Unknown, unknown, *Fraulein Bohm*, *Mödel*, *Nestler*, *Pohl*, unknown, unknown, *Wurzbach*, unknown, *Meyer*, and *Niess*.

***Hanna Reitsch*, *Professor Georgii*, and *Hans Jacobs*. About 1933.**

early December 1941. These men from *LC-1* and *LC-2* included *Knemeyer*, *Schelp*, *Baeumker*, *Antz*, *Reidenbach*, and so on. Each of us at *DFS* knew, for example, that the projects which we were working on were very well advanced. One such project was a twin fuselage aircraft known as the *DFS 332*, which had a center wing which could be exchanged, and which could be installed at different angles of attack in flight so that we could really check the results of wind tunnels in free-flight at higher speeds and higher *Reynolds* numbers. Then there was our *DFS 346*, a pure research aircraft. War, or no war, powerful men in the *RLM* wanted it built and flown. We ran out of time, but it was later attempted to be built by a French group which I had some contact with at that time of the war. But, at this late stage of the war, there were a lot of such developments, research projects like the *228* and *346*. That is, projects which were simultaneously thought up and construction started.

The *DFS 230A* transport glider designed by *Hans Jacobs* and widely used in German paratroop airborne operations.

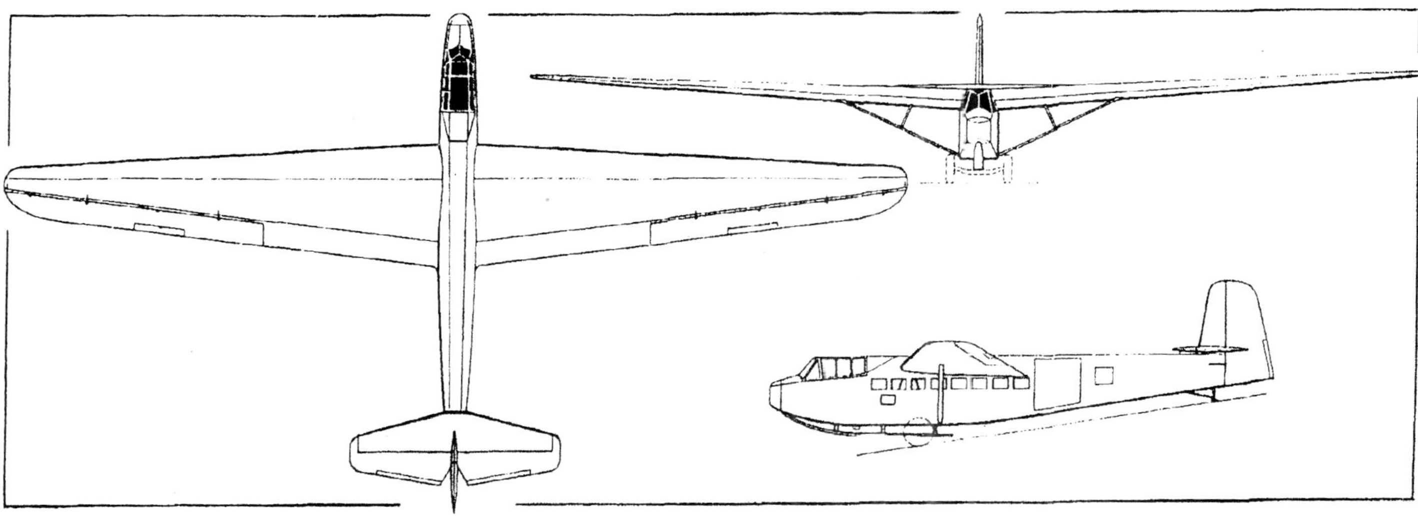

A pen and ink 3-view illustration of *Hans Jacobs' DFS 230A* transport glider.

A starboard side view of the *DFS 230A* cargo/transport glider designed by *Hans Jacobs*.

Hans Niess (sitting on the ground) with *Wolf Hirth*. Hornberg *Flugplatz*, near Schwäbisch Gmünd, West Germany. Summer 1954.

Even at the top level of the *RLM's* General Staff there was a clear feeling about the need for reconnaissance, even this late in the war by those in the military business.

MYHRA - What was the name of this person in *LC-1* at that time? Was it *Hans Antz*?

KRACHT - No, not *Antz*, because he was in *LC-2* or the air frame department. In *LC-1* it was *Dr*...oh, I've forgotten. By the way, *Antz* went to the USA after the war.

MYHRA - What about *Oberst Siegfried Knemeyer*?

KRACHT - Ah, *Knemeyer*, a very curious man and a very intelligent man, too, he was. *Knemeyer* was also one of those powerful men in the *RLM* and, as far as I'm concerned, its unofficial leader. *Knemeyer* had been at the Erprobungsstelle-Rechlin (Aircraft Testing Center-Rechlin), and then later on in did photo reconnaissance all around the world, for which he had been awarded the *Ritterkreuz*.

MYHRA - In late 1943 or early 1944, *Knemeyer* took over *Udet's* old job from *Milch*. He came to be the director of the *RLM's Technical Development*, and perhaps *Göring's* most trusted advisor.

KRACHT - I don't remember this, but *Knemeyer* was really linked to everyone in the *Luftwaffe*. The name of *Knemeyer* had been known by all the pilots because he had developed a navigational computer which was simply called the "*Knemeyer*." But the real man was known only to a very few people. He was involved in a lot of things. All these people, I think, now give historians the idea that the *RLM*

Hans Jacobs at his home in Siegen, West Germany. 15 May 1987. Photo by the author.

Neiss (left) and *Mödel* playing in the snow. *DFS*-Ainring. Winter 1942.

***Neiss*, facing the camera. *DFS*-Ainring. Winter 1942.**

***DFS*-Ainring's winter experimental station where *Felix Kracht* and others conducted tests on the prototype *DFS 54's* pressurized cockpit cabin.**

was a rigid organization, when in fact it was very fluid and penetrating from one side to the other. If there was something of interest, then there were no walls between the several departments. If some interesting thing appeared on one side, immediately individuals on the other side were involved, and this explains the fast rise and high-speed development of the aircraft industry in Germany from 1933-1945. Otherwise, our rapid aeronautical achievements would not be understandable, that from scratch in 1933, we just developed it into one of the best air forces and aircraft industry that the world had ever seen.

I recall that when the first *North American P-51 "Mustang"* was shot down it came down not too badly damaged. It was dismantled, and its wing was immediately taken to the big wind tunnel at *AVA*-Berlin/Adlershof and tested there in its wind tunnel. This was the first thing we wanted to do with it, even before anyone thought about flying the enemy fighter, because it was known that

Members of the *DFS*-Ainring's *DFS 54* group on a mountain climbing venture.

A pen and ink illustration of the *DFS*-Ainring reworked *Me 328 Schmidt-Argus 014* pulse-jet powered fighter interceptor. *Niess* was the design chief on this ill-fated project taken over from *Messerschmitt AG*.

A pen and ink illustration featuring the starboard side of the *DFS*-Ainring reworked *Me 328 Schmidt-Angus 014* pulse-jet powered fighter interceptor in gliding flight.

The prototype *Me 328* at *DFS*-Ainring mounted on a *DFS Do 217E* carrier aircraft used for extensive gliding trials. *DFS* was looking into how these wood and steel pulse-jet fighter/interceptors could be mass produced in the most efficient manner.

the "*Mustang*" had a "laminar" flow airfoil. We also knew about laminar flow, but we had not yet applied it to our own aircraft. Constructing a laminar flow wing was a difficult industrial process that our industry could not give at the time. Plus, under wartime conditions, the tolerances needed for its manufacture required the use of pure sheet metal. It was just not possible. Today, of course, laminar flow construction is quite common. On the other hand, with the development of the "*Spitfire*," its wing construction was quite standard. We rechecked these aircraft wings against *Bf 109s*, and thus we could clearly state that at "X" altitude the "*Spitfire*" was likely to be better than the *Bf 109*, and these altitudes should be avoided. In other words, go higher or lower, but stay out of "V" altitude. So, all this wing stuff was clearly understood because we had flown every Allied aircraft.

MYHRA - Yes, there was a *Luftwaffe* pilot by the name of *Theodor Rosarious* who flew many Allied aircraft in flying demonstrations against *Bf 109s* and *Fw 190s*. I've met with him several times.

KRACHT - Yes, I knew him. He was a fantastic man, but very often for him nothing was impossible. I told you that the origins of *DFS* came about through sailplane research. But getting back, if you look into the combination of towing aircraft, towing aircraft, and towed aircraft, you will see that the towed aircraft suffers the rough air flowing back from the towing aircraft. It means that the performance of the whole is worse than the performance of the two taken individually. We had in the *RLM's* General Staff a great deal of interest about refueling, but also of saving aircraft in flight, which was experiencing difficulty, that is, aircraft experiencing some trouble such as with engines or whatever. We all wanted to know if it was possible to recover an aircraft in flight and bring it back. This was particularly true for the *Fw 200* "*Condor*," our long range reconnaissance aircraft. The *RLM* asked us if there was any possibility to bring back an aircraft which was experiencing difficulty which

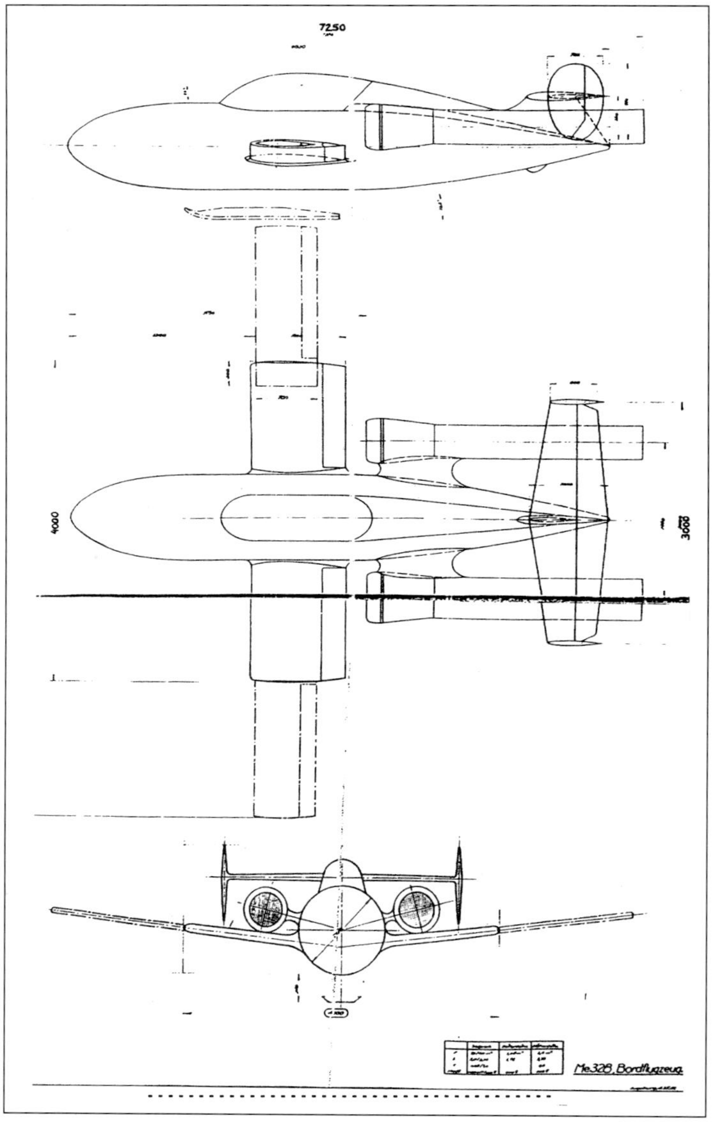

A pen and ink 3-view drawing of the *Me 328B* from *Messerschmitt AG*. *DFS*-Ainring. Under the leadership of *Niess*, three wood and steel prototypes were constructed by *DFS*.

A scale model of *DFS*-Ainring's final reworked *Me 328B* series production version. *DFS* calculated that these parasite pulse-jet powered fighter/interceptors could be constructed in only 4,200 man hours.

would otherwise have to be ditched in the sea? About this time, too, the idea came down from the *RLM* asking whether it would be possible to tow another aircraft like a fighter on the wing tips of an *Fw 200?* The idea was to couple, say, two fighters together wing tip to wing tip. We had made some tests with this idea and had linked two small two seater aircraft in the air while flying.

MYHRA - What year were these coupling tests made?

KRACHT - This was in 1943-1944, at *DFS*-Ainring near Salzburg. I personally flew one of these aircraft, and a friend of mine flew the other. We tried these experiences with about twenty pilots, and finally we learned that it could be done, because my friend and I proved it to ourselves. Nevertheless, we were very nervous when you went over from normal flying to coupled flight. The process went like this: First, one let out a long wire cable of about 30 meters [96.8 feet], and then the other aircraft hooked it. Second, you manually reduced (pulled it in) to make a fixed coupling. This coupling made you rigid to a point in front, and in the rear the half points were just lying there together. But you had complete freedom to make your own incidence. Then finally we tried it in the air. We found that the ailerons were not working out as planned. On each aircraft its ailerons would just produce bending moments on the wing structure and nothing else. In order to turn you had to accelerate, then with one elevator going down and the other going up, one each against the other, it worked. Also at this time we had no voice recorders and no direct communication, because these aircraft were too small for that, so we communicated through hand signals.

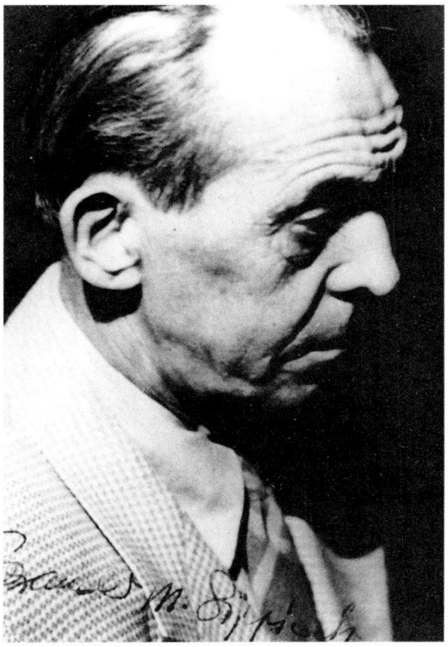

***Dr.-Ing. Alexander Lippisch.* Reunion at the *Rhön/ Wasserkuppe,* February 1973. *Lippisch* had been a senior official at the *DFS* with more experience in rocket propelled aircraft than any other individual in Germany.**

MYHRA - What about the release mechanism in the coupling...how did that function?

KRACHT - This was a simple mechanical coupling. You could free yourself from the other aircraft immediately. This was no problem. With this coupling we had even tried landing together on the ground. It was possible.

MYHRA - What type of aircraft did you use during the coupling experiment?

KRACHT - These were two different type of aircraft. Both were made by *Hans Klemm*, though different models. We tried this coupling experiment first with smaller aircraft, and then eventually with larger aircraft. At *DFS*-Ainring we had a small airfield which was considered *Hitler's* personal airfield. This airfield had no hard surface runway, but one of grass which was well maintained. It was

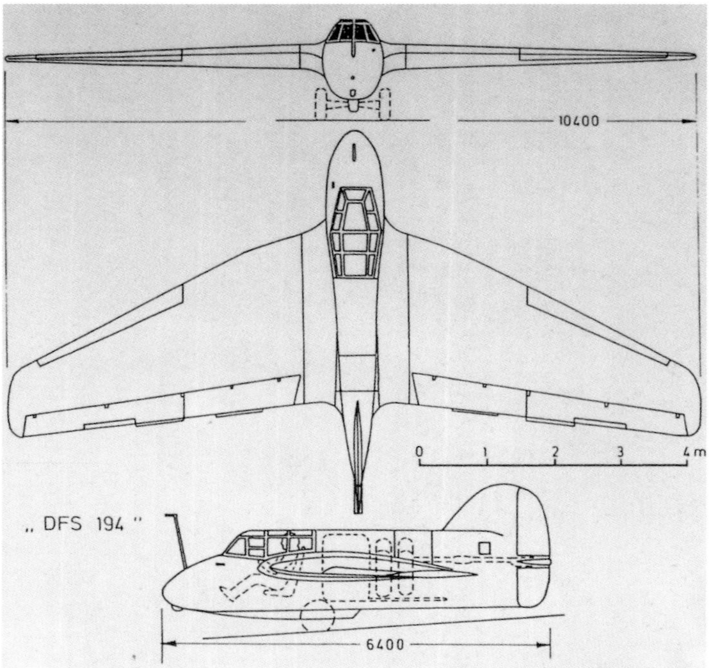

The *DFS 194* by *Alexander Lippisch*. *Lippisch* and he had been associated with *DFS* almost from its very beginning in late 1924 , when it was known as the *Rhön-Rositten-Gesellschaft* (*RRG*). *Lippisch* became the director of the Aeronautical Department in early 1925, and left in 1939, to join *Messerschmitt AG* in the series production of his bi-fuel liquid rocket-powered machine, which had become designated the *Me 163*. *Lippisch* didn't have any meaningful involvement in either the *DFS 228* or *346*.

located about 500 meters [1,640 feet] west of the border between Austria and Germany, so when you came in for landing you had to fly over Austrian territory. But, during the war this was no problem.

MYHRA - I imagine that coupled aircraft were very difficult to fly?

KRACHT - No, not really. It was easy to fly if you let the aircraft go. The difficulty came when the pilot wanted to use too much control and over controlled. There was one experience when we had made a receiving cone just 1 meter or 0.50 meters distance. Not high, not low, not to one side, but just behind the aircraft. This was really a curious experience for most pilots, and most of them overreacted. I had one experience with *Hanna Reitsch* who said that she was not moving... it was the coupling cone that was going around. *Reitsch* said she didn't want to do those coupling experiments again.

MYHRA - What I don't understand is this: given the time when the *DFS 346's* design was started, in early 1944, what was its stated mission, and what was the need or reasoning behind it at that late hour of the war?

KRACHT - At this time—around early 1945—no one was thinking about an aircraft for the Front anymore. You could see the war effort being supported by America and what was happening to our homeland due to the USAAF's bombing. Despite this hardship you had a task. You did it, to make it go no matter what was happening around you. If you didn't you'd likely end up on the *Russian Front*. So we did what we were told to do. I'll give you another example. In the spring of 1944, I was charged, in addition to my other tasks, such as coupling and in flight refueling at *DFS*, to take over the management of a small workshop which would manufacture aircraft wings. This workshop had been closed at the beginning of the war due to the fact that it formerly had manufactured consumer goods of some kind—furniture, I believe. So I received an order to reopen this workshop and to manufacture wings out of wood for

Wing tip to wing tip coupling experiments conducted by *Felix Kracht*. The idea was to attach one aircraft to the other in cases of long range flying or when one aircraft was experiencing motor trouble and the other aircraft could connect and bring it back to the air field.

Here the two early model *Focke-Wulf Fw 56* tandem trainers have joined together. *Felix Kracht*, himself, was one of the pilots, and *Karl Schieferstein* was the other.

the *Heinkel He 162 "Volksjager*." This aircraft had just made its first flight in March 1944. Then the order came to build 1,500 of these *He 162s*. With the order I reopened this factory in May 1944 to July 1944, and I delivered wings for the *He 162* each week . If anyone speaks of efficiency, I then can certainly refer to the production of wings at this old factory. I was ordered, too, to go after all the other parts needed for the assembly of the *He 162*, which due to the disruption caused by the war was keeping the production of 1,500 *He 162s* unfilled. I was now driving 200 to 300 kilometers [124 to 186 miles] per day seeing to it that the necessary parts and equipment were being produced and delivered.

MYHRA - Where was your factory located, and who gave the order to produce these wings?

The coupled *Focke-Wulf* aircraft as seen from below.

***DFS* test pilot *Karl Schieferstein*. September 1942.**

Karl Schieferstein. **Königswinter bei Bonn, West Gernany. 1984. Photo by the author.**

KRACHT - It was located next to our *DFS* facility in Ainring, about 2 miles north of Ainring near a railroad station close to the border between Austria and Germany. It was an assignment/order from the *RLM*.

MYHRA - Where did the idea for the *DFS 228* come from?

KRACHT - The *DFS 228* was not the idea of a single person. It was a combined idea of people from the department of research and development, or *LC-1*, in the *RLM*. There came to *LC-1*'s attention suggestions from their own people, from pure science at the universities, suggestions from the aircraft industry, from government aeronautical research centers, wind tunnel people, and the like, and this is how the project idea came about.

MYHRA - How much input did *Professor Georgii* have on its design and production?

KRACHT - None. None at all. He wasn't involved, for at this time he was the top manager of *DFS* and was involved in many, many aviation related activities of the organization. *Georgii* did not interfere in the design details...no, not at all, and especially so on rockets. He had no interest in getting involved in rocket-powered aircraft such as the *DFS 228* and *DFS 346*. I really didn't care, because I had sufficient work to do with the *DFS 228* project, the "father and son" project (*Ju 88* and *Bf 109*), and other things. We at *DFS* were doing a lot of things for the *Junkers Flugzeugbau*, and we had been flight testing the *Me 328*. You know, that small aircraft with twin pulse-jet type engines. Simultaneously, my small factory, which had been put in charge of building the wings for the *He 162* "*Volksjäger*, did not get the assignment after all to help construct the Me 328 series. The project was caceled. But we never did complain that we had no work to do. We had plenty.

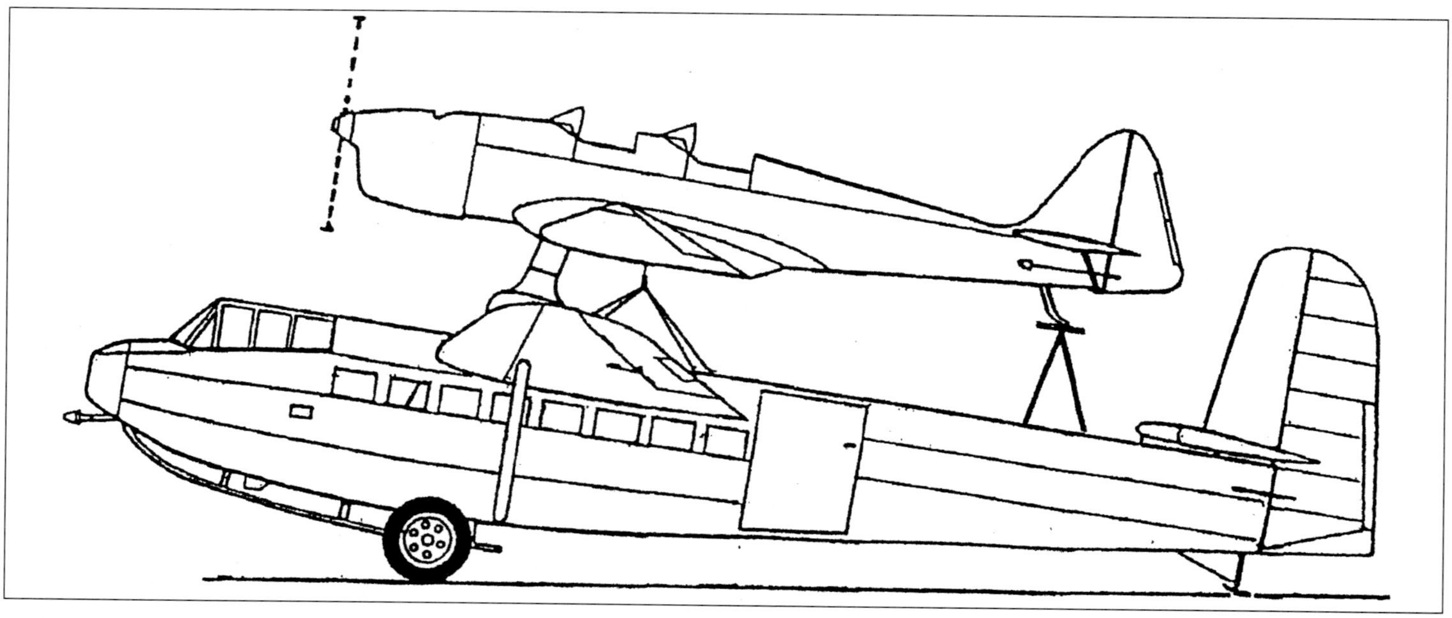

A pen and ink drawing featuring the initial "*Mistel*" experiments at *DFS*-Ainring. Shown is a *Klemm Kl 35* atop a *DFS 230* transport glider. The *DFS 230* was designed by *Hans Jacobs* of *DFS*.

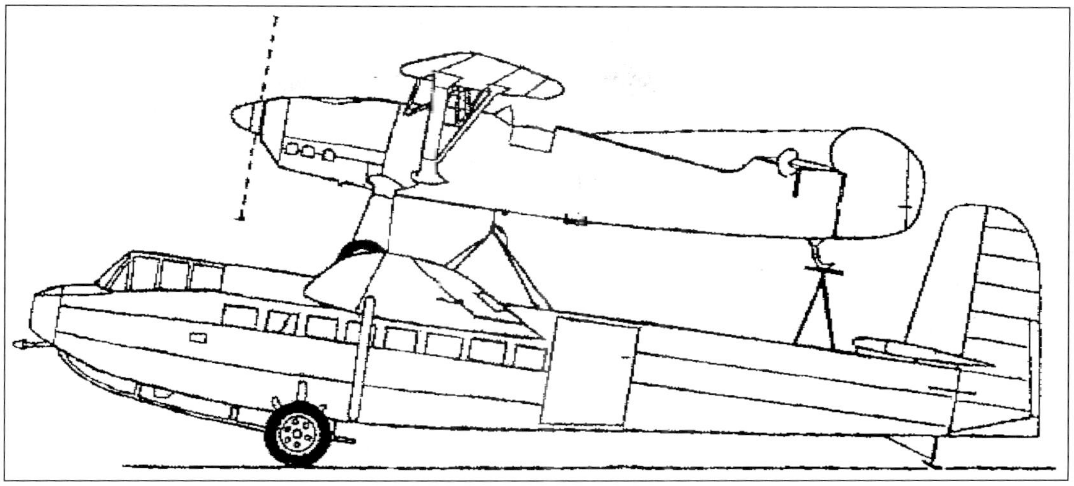

A pen and ink drawing of the second "*Mistel*" experiments this time featuring a *Focke-Wulf Fw 56 "Stösser"* atop a *DFS 230* transport glider. *DFS*-Ainring. September 1942.

MYHRA - Tell me about yourself...your education and aviation experience.

KRACHT - Oh, this is very simple. I started my aeronautical studies at the University of Aachen under *Professor Theodore von Karmán*. He was a professor of aerodynamics there. He also was probably the only American that's ever been a member of a Communist government. After *WW1,* he was a member of the Hungarian government, and this was a Communist government. *Von Karmán* usually traveled to the USA two or three times a year. Then he immigrated to the USA in 1932, but he was able to maintain his chair of aeronautics at the University right up the beginning of the war in 1939. He was Jewish, too.

MYHRA - How was he able to maintain his chair when he lived in the USA and was a Jew?

KRACHT - Look, at this time there was an unofficial saying from *Göring*. He said that he alone in the *Third Reich* decided who was or was not a Jew.

MYHRA - I thought that *Göring* was only referring to *Milch*, This included *von Karmán*, too?

KRACHT - No, not *von Karmán*, but there was certainly more behind it. *Von Karmán* was what people would call a "double-agent."

MYHRA - Well, I've never heard this allegation before. He was, however, a curious person in the USA. He liked to associate with Hollywood female movie stars...those so-called sex queens of the late 1940s and 1950s, such as *Mamie van Doren*, *Jayne Mansfield*, and others.

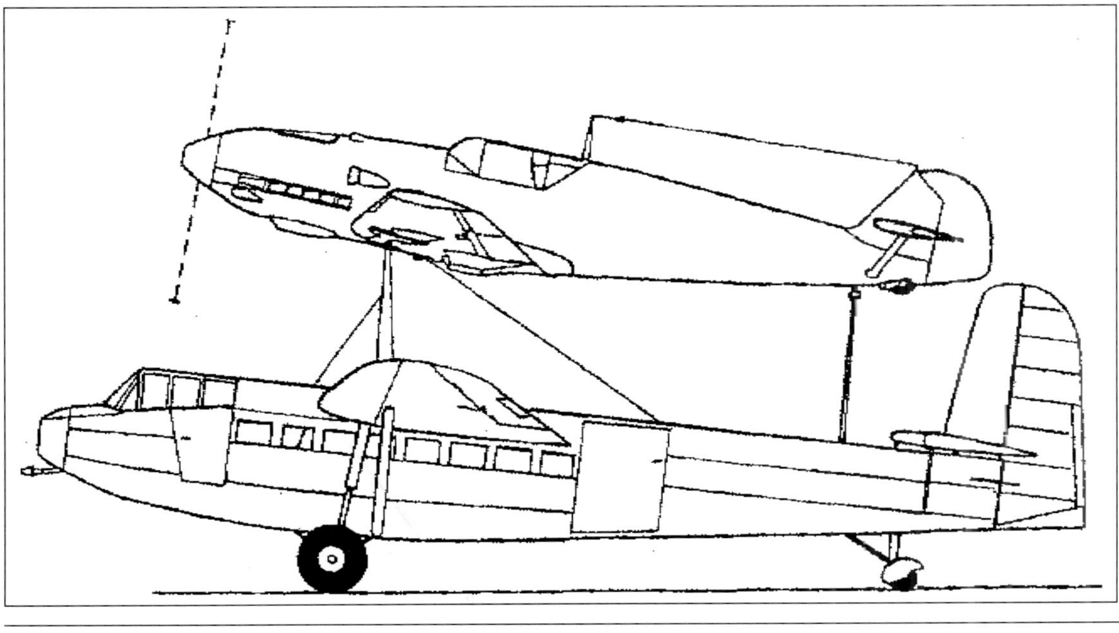

A pen and ink drawing of the third "*Mistel*" experiments this time featuring a *Messerschmitt Bf 109E* atop a *DFS 230* transport glider. *DFS*-Ainring. June 1943.

DFS **test pilot** ***Karl Schieferstein*** **piloting the** ***Bf 109E*** **control aircraft atop a** ***DFS 230.*** **Hörsching near Linz. 21 June 1943.**

KRACHT - He was a "ladies man" with his female students, too. But we suspected that when he was at the University of Aachen that he was involved in something else. He had many girlfriends at the University. He was a genuine person on the scientific side and on the financial side. He was able to build up his check book. He was a very clever man, too. He was an Austrian officer in *WW1*. In 1918, he became a minister in the first Hungarian government after the war. This was a Communist government, and they were in a strong position party in Hungary. Then *Theodore von Karmán* came to Göttingen, and then he started his scientific career. I finished my academic studies in 1938, and during these studies I had had very good training as a pilot. At this time the *RLM* had selected in the different universities some twenty students who had the qualifications for being a good pilot, and then they trained them as future test pilots (*Flugbaumeister*). At the same time I had done a lot of soaring and powered flights. I had made my first crossing of the

Felix Kracht's **experiments in rigid towing of one aircraft by the other. Shown is an aircraft easing its way up to the towing aircraft...a** ***Ju 52/3m.***

DFS **test pilot** ***Karl Schieferstein*** **testing the "*Mistel*" concept. He is piloting the** ***Bf 109E*** **control aircraft atop a** ***Ju 88.***

A view from the cockpit of a *DFS* aircraft on a tow line to the *Ju 52/3m*. The pilot is attempting to rigidly attach itself to the leading aircraft by drawing in the tow line.

From the cockpit of the towed aircraft it appears that the machine is about to make contact with the *Ju 52/3m* and form a hookup.

Alps from Augsburg to the Italian plains in a sailplane that I had designed during my studies in Aachen. When I finished school I sold the sailplane. Then came the war, and all of Germany was finished with sailplanes, so I came at that time to the *DFS*, which offered me the job as director of the aircraft institute department of *DFS*.

Getting back to my aviation experience, I have flown extensively, most aircraft built in Germany up to *WW2*, beginning with gliders, including the *Me 262*. I had a very hard time, because when I started in 1927-1928 as a boy of 16, the left bank of the Rhine where I was living was always occupied by the Belgium and British troops, and any flying by us Germans was forbidden. So I had to go on the right side of the Rhine to an aero club. This was quite an experience, because on weekends we had to travel some 60 miles [95.6 kilometers] by bicycle. I obtained my glider license when I was 16 years old. Then, after a time, I received real training, and later when I went to the Technical School of Aachen, then my aviation activities started up again very quickly. There we flew aircraft which we designed and built ourselves. In this way, I was responsible for the design of an aircraft. It was nicknamed the "*Rheinland*." With this aircraft I participated in the first flying experience in the high mountains of upper Bavaria to Austria. These flights gave us a lot of experience, and this was in 1936-1937, before the occupation of Austria by Germany. Prior to the occupation, Germany's attitude toward Austria virtually killed the tourism industry, which had been their main source of income. The German government instituted a regulation that you had to have a visa to go out of Germany into Austria. For a visa at this time you had to pay 1,000 Reich*marks*.

Featured in this photograph is a *DFS 230* transport glider hooked up to the tail of the *Ju 52/3m* towing aircraft. This form of aerial towing was known at *DFS*-Ainring as "*Starrschlepp*" or rigid towing.

The nose of an aircraft to be towed via a rigid mechanism with its towing bar installed.

The attachment head, uncovered, of the mechanism carried by the towed aircraft when it joins up with the towing aircraft.

So only a very, very few people could go there anymore. But the sailplane clubs in different German cities were supported by the *RLM*, which had an interest in developing aviation engineers. So, we got the authorization to go out of Germany. The back side of the story was that we had the authorization, and every time we flew into Austria and landed our sailplanes there we were surrounded by angry Austrian people. We had a procedure. We would leave the sailplane where it landed and then go to a nearby telephone to announce back home that we had landed in Austria. One particular time, after making the telephone call, I went back to my sailplane and was promptly arrested and placed in jail. I had verified to the authorities who I was and what I was doing, and I told them that they had to handle us *RLM* sponsored sailplane pilots with care. I knew a lot of men who landed in Austria and who were placed in jail for a short time. It was quite a good experience for us. So, in 1937, I made my first flight over the Alps, as I had had this hope, since a boy, to do this feat. In 1936, I went to the Paris Air Show, and to go from Aachen to Paris took about 3 hours by car. Then it took us months, in fact, about 9 months of preparation. Most people don't recall much about this time anymore, but to go to Paris we first had to have a German passport. At this time to obtain a passport if you were within the normal age of military service was not easy.

MYHRA - All of this because of Germany's remilitarization?

KRACHT - Yes, this was the time of the remilitarization of Germany. To go abroad was not so easy as a young boy within the age of normal military service. They checked first to see if you had any reason to escape Germany, and then with your passport in hand you had to go next to Dusseldorf in order to obtain an exit visa. You paid a fee for the visa. But every person had to present himself their

A close-up view of *Felix Kracht's* attaching/holding mechanism used in rigidly towing an aircraft. The towed aircraft carries this mechanism.

The entire rigid towing mechanism developed by *Felix Kracht*. This (male) connector bar, with its attachment seen to the left in the photo, would be fixed within the nose of the towed aircraft. The metal covering over its head has been removed from this mechanism.

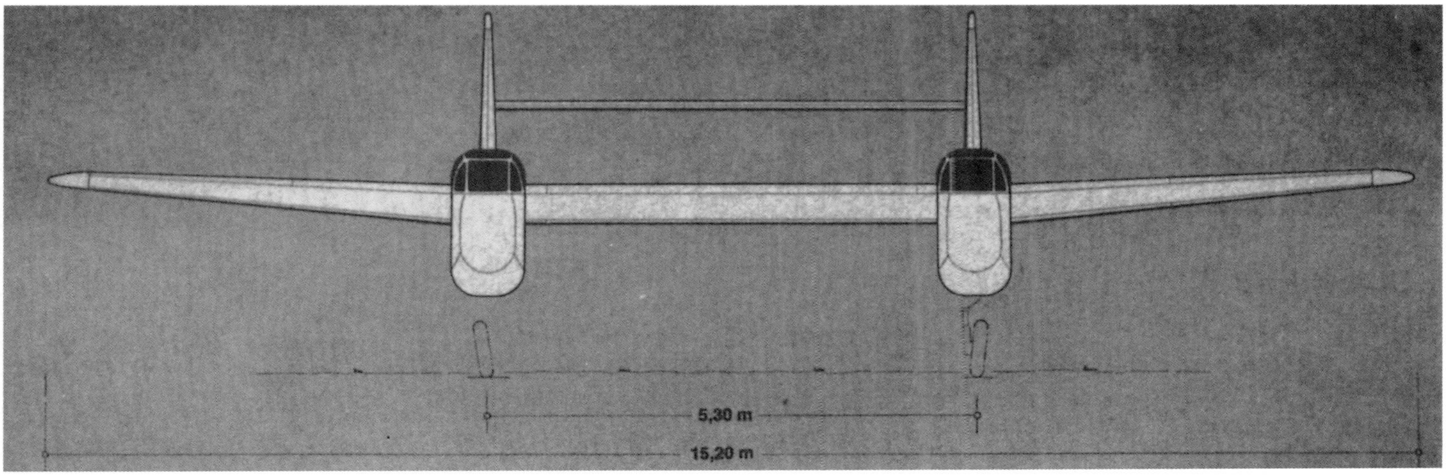

A pen and ink drawing of the proposed *DFS 332* featuring a ground level view.

in person for the visa. It meant, in our group, that no one person could collect all our passports and take them to Dusseldorf and get a visa stamp. Each passport holder had to go personally to the Belgium Embassy. Then, two to three weeks later, you had an invitation to appear a second time in Dusseldorf, this time at the French Embassy. Thus, we had to go back to Dusseldorf again, because Belgium had given us a visa to go from Germany to France, but not back from France to Germany. They would not take our money until we had received the visa from the French. This is why it took us nine months for all the paper work, authorization, and to pay the fees. Plus, we had to exchange *Reichmarks* into *French Francs*. In addition, it was unthinkable to go by auto. We took the train. Today, if you go across the border, maybe a border guard will ask to check your passport, maybe not. Anyway, before I had finished my stud-

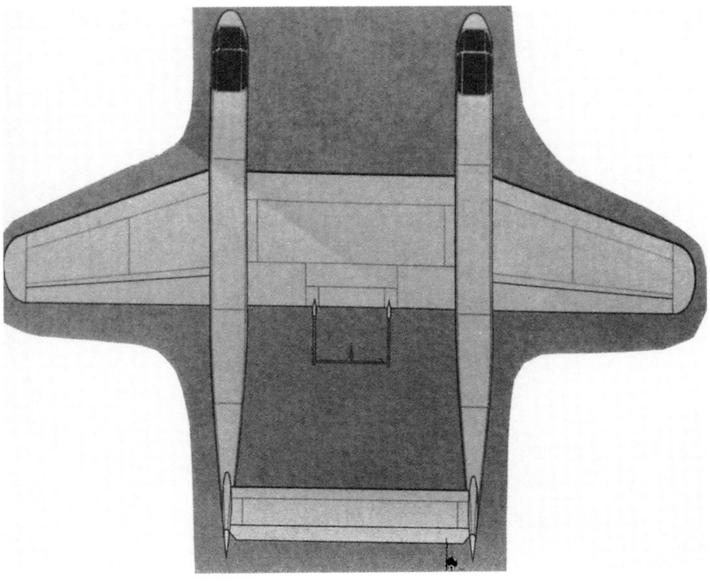

A pen and ink general arrangement view of the proposed *DFS 332*.

***Werner Blech* shown in the cockpit of his *Horten Ho 3c* at the August 1938, Summer Sailplane Competitions at the *Rhön/Wasserkuppe*. 6 August 1938.**

***Werner Blech*, and his *Horten Ho 3c* with its forewing, waiting for his turn to be towed up into the skies at the 1938 Summer Sailplane Competitions, *Rhön/Wasserkuppe*. 6 August 1938.**

***Werner Blech*, seen overhead, piloting his *Horten Ho 3c*, civil registration *D-12-347*, at the 1938 Summer Sailplane Competitions. *Rhön/Wasserkuppe*. 6 August 1938.**

ies on aviation at Aachen, I and a friend made up a small shop, and up to the beginning of the war we had built some 50 glider aircraft. Two or three are still flyable today.

MYHRA - And the name of these flying machines?

KRACHT - They are all called the "*Rheinland*." When war started, all soaring and sailplane activities were interrupted, and therefore, I looked around for work and then accepted an offer from *Professor Georgii* to take over the aircraft department in the *DFS*.

MYHRA - Did you ever earn a doctor's degree?

KRACHT - No, I never pursued a doctor's degree. Instead, I have the Dipl.-Ing. (*Bachelor's Degree*). This I have. I had been invited several times during the war to present one of my papers to become a doctor, but for personal reasons I always refused. Because, essentially, it was a kind of pressure by the *Nazi Partei*, that you could have a doctorate if simultaneously you entered the *Nazi Partei*. This I would not do.

MYHRA - *Lippisch* obtained his doctorate in 1943, apparently by merely presenting one of his papers.

KRACHT - Yes, I know. With a doctorate you would simultaneously have been given the rank of *Professor*. All this was a normal procedure to put pressure on people who otherwise would not have been attracted to the *Nazi Partei*, and the *Partei* needed people like *Lippisch*.

MYHRA - I am not aware that *Lippisch* was a *Nazi Partei* member. *Willy Messerschmitt* accepted a position of *Professor*, too.

The hail-stoned damaged port outer wing of *Werner Blech*'s *Horten Ho 3c* at the 1938 Summer Sailplane Competitions. *Rhön/Wasserkuppe*. 6 August 1938.

The upper side of the sheared off port outer wing tip of *Werner Blech's Horten Ho 3c* all-wing sailplane.

***Horten Flugzeugbau* 1938 Summer Sailplane Competitions team member *Heinz Scheidhauer*.**

A poor quality photograph of *Heinz Scheidhauer's Horten Ho 3a,* civil registration *D-12-348*, at the 1938 Summer Sailplane Competitions. *Scheidhauer* can be seen standing up in his cockpit.

KRACHT - Oh, but very late. He was given the rank of *Professor*, but he never gave lectures at a University. It was sort of an honorary title, and in addition, it was given to *Errnst Heinkel*, *Kurt Tank*, and others in the aviation industry. It was a political thing, and they gave them out as a civil award for working hard in aircraft armaments.

MYHRA - When you were working on your *Dip.-Ing.*, did you specialize in a certain part of aviation?

KRACHT - Yes, at this time I specialized essentially in aircraft project designs. That is, complete aircraft projects from idea to full plan definition. This includes the computation of performance, including the computation of the main elements of the aircraft and the aerodynamics, too. This is one of the reasons why I was given the responsibility for the initial design phase of the *DFS 228* and *DFS 346,* as well, because I had a lot of experience and, but curiously I had no experience in production. I had had experience in project management, but not in production. Then, with the war such as it was, I had to take over the responsibility for production. The last 35 years now, I have gained experience in production. The last job of my career was director of the French effort on the Air-Bus program. So, I never finished my studies. I joined *DFS* at the end of 1939 and stayed with them until May 1945.

MYHRA - What did you do after May 1945?

KRACHT - I lived in Upper Bavaria from Summer 1945 to Summer 1945, and during this time I thought a long time about what I should do with my life. At this time everyone was sort of fixed to his town or village, and there were few possibilities to move without authorization from the military government. You just couldn't merely move. One day in Summer 1946, I took my bicycle in the direction of Frankfurt where I had some relatives, and I could stay overnight because you couldn't sleep outside in the open air because there were curfews. From there I traveled around all together

The wreckage of *Scheidhauer's Ho 3a* in the trees at the *Rhön/ Wasserkuppe*. 6 August 1938.

***Scheidhauer's Ho 3a* crashed into the trees upside down as shown in the photograph. What is seen is the *Ho 3a's* underside starboard wing. 6 August 1938.**

Shown in this photo are the mangled remains of *Scheidhauer's Ho 3a* port side outer wing. 6 August 1938.

***Heinz Scheidhauer*, Staufen/Breisgau, West Germany, July 1988. Photo by the author.**

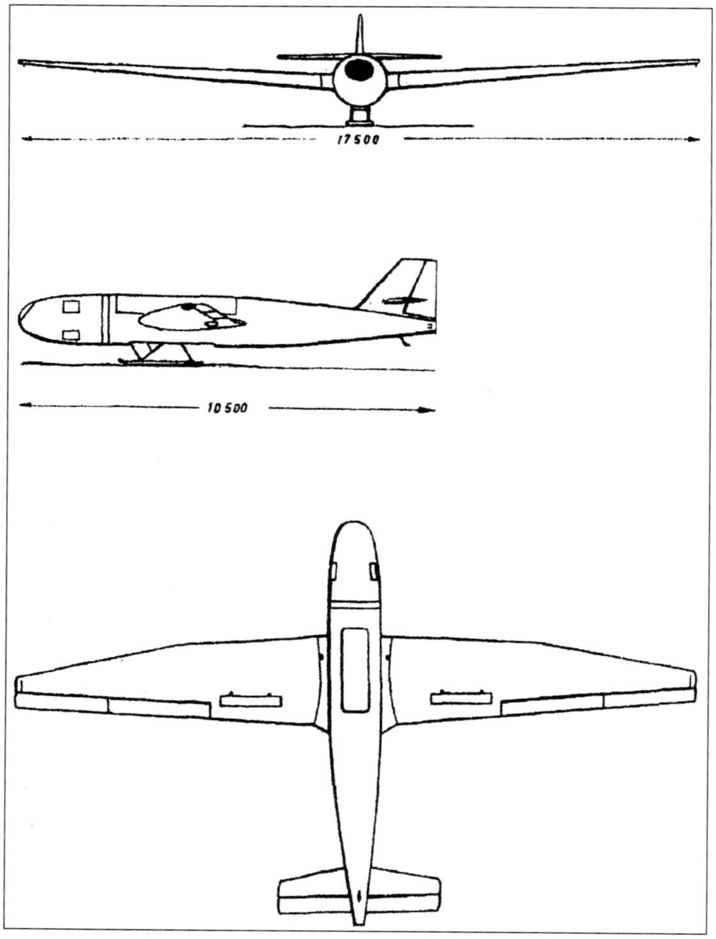

A pen and ink 3-view drawing of *Felix Kracht's DFS 228 V1*. Its wing span was 57 feet 1/2 inches, and its overall length was 34 feet 8 1/2 inches.

about seven weeks, but it was useless, for nothing remained of the aircraft business. Working relationships even between the British, Americans, and French were not that easy, either. I realized that in the immediate future nothing in terms of aviation flying and design would happen in Germany. The only problem was to survive from day to day. So, with the few possibilities that remained, I started up a small business with the people who stayed with us. It was a shoe repair business. We German people had many, many things which needed repair. This helped me to survive. Then, in Spring 1946, I soon had employment inquires from American aviation companies. One was from *Lockheed*, but they didn't work out. Then I went to the French authorities and talked about my high-speed aviation work at DFS. So, in June 1946, the French authorities in Germany proposed that I go to Paris and work for the French aviation industry. I told them, look, I'm not just interested in having a job, but I'm interested in a certain job, and I'd like to know what I have to do? Finally, I reached an agreement with the French, and they flew me to Paris in an old *Junkers Ju 52/3m*. In Paris I found there a man which I had known during the war, and who was now with the French aircraft establishment, and he proposed that I should work with him on some high-speed aircraft. Finally, after a lot of thinking, I accepted. So now, in the Summer of 1946, I was firmly established in France. The French were basing their aircraft industry after the war at least on some of the German turbojet engines, such as the *BMW 003*A and the *Junkers Jumo 004B*. But the French needed spare parts for these turbines and so on. They did not know how to locate these parts in Germany, so I traveled extensively throughout

Germany looking for someone to make them. So, from the early beginning in France I had to help them maintain some of our ex-German aircraft and engines. I developed a lot of contacts with German industries, and in the meantime I developed a lot of contacts with the French aviation industry, too. Also, flight test equipment was required, and then in 1955, when the German aircraft industry was restarted, the French asked me to be one of their representatives. I worked for a company at the time, and we started production of a transport aircraft in Germany under license. This then led, in the 1960s, to the joint planning and development of the supersonic *Concorde* aircraft, which I was involved in, especially with its serial production. Then came the Air-Bus program, and I continued to have my office in Paris but also lived in Germany. So, for forty years I have been living in France and West Germany.

MYHRA - Prior to joining *DFS* in 1939, did you talk to any other aviation firms about employment, such as *Heinkel AG* or *Messerschmitt AG*?

KRACHT - I had, at this moment, offers from all of them, but finally I chose to join *DFS*. I had met, too, the head of *Blohm & Voss*' aircraft design group, *Dr.-Ing. Richard Vogt*.

MYHRA - *DFS*, in 1939-1940, had lost *Lippisch* and *Jacobs*. The aviation side of the society must have reached a low point in its history?

KRACHT - No. You have to know these people. *Lippisch* always in his life has been a genius in new techniques, and he has had many good ideas. But the real genius, in my mind, is the person who can take the idea and make it work. *Lippisch* initially came to *DFS* to transform ideas into hardware, but he was not interested in this next step, because by that time he had already gone on to his next design idea. This is why, in the career of *Lippisch,* you have such ups and downs. For example, when he had developed the tailless *DFS 194*—by the way the only project of his, in his whole life, which had really come to be—had to be built by Me*sserschmitt AG*. So, this is why he had left *DFS*. On the other hand, *Jacobs* was into transport gliders, I believe, and joined *Gothaer*-Gotha.

MYHRA - What kind of man was *Hans Jacobs*?

KRACHT - He was actually my predecessor. *Jacobs* had the institute for aircraft at the beginning, and then in 1940, he left *DFS* to join *Gothaer* and developed there another military glider. He stayed in the German aviation industry throughout the war, I think forming his own firm late in the war.

MYHRA - Why did *Lippisch* and *Jacobs* both leave *DFS* for the aviation industry? What were the reasons?

KRACHT - I don't recall. I assume that it was for personal reasons. At this time, despite what we generally thought, there was no pressure on them to stay at *DFS* as long as they remained in the general aviation business. The German aircraft industry was undergoing a tremendous amount of change at this time, and people were shifting back and forth. No, there was no low point with their leaving, because *DFS* had other activities, too, in aerodynamics, and in particular, the aerodynamics associated with the development of the first turbojet engine. The theory, or rule, about air intakes, which is even valid today, had been developed at the *DFS* by *Professor Ruben*, *Professor Flettner*, and others. But this work of *DFS* specialists is completely unknown. There is one man who knows more about *DFS* than any other alive today. His name is *Karl Schieferstein*. I will give you one other name, and he was one of the main pilots at *DFS*. His name is *Hermann Zitter*. Each man started out at *DFS* as pilots from practically the beginning of *DFS*, and as such they have had a bit more time to think about the history of the *DFS* society.

MYHRA - Tell me about *DFS*' director, *Professor Georgii*.

KRACHT - *Georgii* was a meteorologist and was very early in favor of the first expositions of sailplanes to study the atmosphere. He was appointed to the American Meteorologist Society in Germany. At this time he was living in Hamburg. On his recommendation, two sailplane societies, the Rhön Society and the *Rossitten Society*, were merged, forming the *Rhön-Rositten Gesellschaft (RRG)*. Later on, after the merger, about the time the *RLM* had been created in 1933, *Georgii* was invited to present a report to the *RLM* on the work done by the RRG. He stated in his report that they were not doing for sailplanes what they had hoped in the beginning. *Georgii* was later invited to take over the *RRG* and transform it into the *DFS*. At the same time, too, they created a chair of meteorology at the University of Darmstadt for him. Then, I think in 1926, *DFS* in Darmstadt was created. It remained there until 1939. At the beginning of the war, however, the German government had been giving *DFS* work which had relatively little to do with gliding and soaring, but in particular work on advanced aerodynamics for military aircraft. Because of the war and *DFS*' military work, the *Reich* government decided that Darmstadt was too close to the German/French border, and *DFS* was moved to Braunschweig from Darmstadt. There it was combined with another research center, *DFR*, and you know when two research centers are sitting together, on the same ground, it does not work. At this time, they looked at other locations. There were disputes with *Steiner*, the right hand man of *Georgii*, and really one of the fathers of soaring. Because of *Steiner* we had been working together with the *RLM*, and we should accept the offer to go to Magdeburg. There was a *Luftwaffe* base at Magdeburg which was practically empty because its people were at the *Front*, and this location had been offered to us. Traveling time between Magdeburg to Berlin was only two hours by train, so the links would be easy. Yet, I opposed this due to the human factor. I said no, we should instead go to Ainring. There was a *Luftwaffe* base there, too, and it was also empty. But no one would come to Magdeburg from Berlin. If we were sitting in Ainring, then we would have to make up a waiting list of the people we would accept to come, because Ainring is a nice mountainous country and people can go skiing, hiking in the mountains, and so on. It would be highly popular. Thus, we could have the people in Ainring we wish to have, and it would give us good public relations in the *RLM*. Finally, *Georgii* accepted my argument, and we moved to Ainring. As it turned out, we had more visitors than we wanted. I was among the first people to move to Ainring. I was never with *DFS*-Braunschweig. I had been with *DFS*-Darmstadt for some high performance pilot training in 1935, to supplement our training in normal flying. I was there with a small selection of pilots which had at their disposal most of all the best high performance gliders that then existed in Germany at that time. This was one of the different steps of selection by the *RLM* leading up to the title of "*Flugbaumeister*."

MYHRA - You must have a great deal of flying experience. How were you able to gain all this flight experience?

KRACHT - Yes. I've tested many, many German aircraft, and have flown a lot of aircraft types. When the student-pilots had earned their diploma in aviation, about 10 of them per year were invited to join a *DVL* training center at Berlin/Adlershof. This was a two year supplemental program, and ended with the title of "*Flugbaumeister*." As such, it was an administrative position/title, and at this moment of the start of your career, you were all ready to join the upper level management ranks of the *RLM*. *Flugbaumeister* corresponded to a managerial position. In addition to entry into an administrative role in the *RLM*, the *Flugbaumeister* was qualified to join the *RLM's* test centers as test pilots. I was invited in 1938 to become a test pilot, but I refused to do so, because I did not want an administrative career in the *RLM* as a government administrator. I wanted to go industry and stay in aviation-design and equipment. So, I refused to go to the *RLM* or one of its test centers and have never regretted the training that I obtained while becoming a *Flugbaumeister*. While at *DFS*, I was frequently invited to go to *Erprobungsstelle*-Rechlin and to participate in flight testing different German aircraft types. This included many Allied types, too, especially American and British. I've also flown many other aircraft types through my work at the *DFS*.

MYHRA - It was quite an honor to be selected to the program at the *DVL*. Would that later lead to the title of "*Flugbaumeister*?"

KRACHT - Yes. There were only 8 or 10 people selected each year from all of Germany for this program. The fallout was very low. You could say for the two-year program there was one condition which we had to keep. That one requirement was that we candidates could not marry. Unfortunately, out of those 8 or 10 people about 2 per year killed themselves in aircraft accidents. For example, of the 20 pilots that were in our initial *Flugbaumeister* program, three have survived the war, and curiously, they ranked in their class 1st, 2nd, and 3rd. The other 16 had lost their lives during flight testing, while one lost his life fighting on the Front. So, flight testing is a risky business no matter in what country it is done. Yesterday, I told you about the coupling or linking experiences in flight of two aircraft at *DFS*. Well, of all the pilots that we had there, I could do it only with two or three. The others could not because they were too nervous. Even the great *Hanna Reitsch* was not able to couple these *DFS* aircraft together. It was really too hard for almost all pilots.

MYHRA - I recall that *Hans Antz* of the *RLM's LC-1* was a *Flugbaumeister*, too, so not everybody went into test flying, did they?

KRACHT - No. During training, the best pilots were going to *Erprobungsstelle*-Rechlin to become test pilots. The others went into administration at the *RLM's LC-1*, *LC-2*, or other top jobs at the *RLM*. By example, *Hans Sander* was the chief test pilot at *Focke-Wulf*-Bremen, and he was a *Flugbaumeister*, too.

MYHRA - Yes, I am familiar with his flight testing of the wooden *Fw Ta 154* "*Mosquito*" prototype. In addition, there is a man living in the USA who was a *Flugbaumeister*, and he specialized in turbojet development at the *RLM*. His name is *Helmuth Schelp*. When you joined the *DFS*, what did *Georgii* want you to do? What ideas did he have for you to work on? What goals and objectives?

KRACHT - First, there was an immediate requirement from the R*LM* for a double seat glider trainer. This was his first task he gave to me, but it was canceled about three months after I started it.

MYHRA - What would have been the purpose of this double seater?

KRACHT - For pilot training, and to be able to send some of the pilot/teachers back to duty with the *Luftwaffe*, because we had seen later on the difference which existed between the pilots from the USA and the Germans. The American pilot for a *B-24* had not flown any other aircraft after basic piloting training. Thus, he was trained to fly only the *B-24* and to fly it in formation behind a leader. Normally, these American pilots did not even know the navigation routes. They were accustomed to following their leader. It was very well known in Germany later on that when you found a bomber formation you must first knock out its leader. Well, in Germany they found that they could obtain better results in the end by training its pilots on sailplanes.

On the other hand, at this time, the normal life of an aircraft was very short. Our aircraft engines were built to last about 100 flight hours. Today, we have guarantees of 15,000 hours. For the aircraft itself, the duration was between 100 and 200 hours. But for the average *Bf 109*, its life did not exceed 25 flight hours! The average *109* was damaged beyond repair/crashed, or lost on the average of 25 flight hours. Many *109s* crashed while being delivered to the *Luftwaffe* bases after their first takeoff, when the pilot pushed too fast and then got into trouble. This was due to the gyro effect of the propeller, and it would make the aircraft make a left hand turn. Normally one, too, would lose the undercarriage, but I learned later on that the losses with the *Bf 109* were similar to losses of the *Spitfire*.

MYHRA - The *DFS* twin seat glider was to help with training pilots for the *Luftwaffe*?

KRACHT - Yes. But you did not train them on a single seater, where the risk of errors were very high. So, with a double seater you could train them in a few flights and then let them fly, and afterward put them in another, more advanced aircraft. For some reason this project was canceled relatively quickly. At this moment we were in Ainring—it was from mid-summer 1940. I started my next task, and this was to develop a mechanism and procedure for coupling two or more aircraft while in the air.

We had the disaster at the August 1938 *Rhön/Wasserkuppe Summer Sailplane Competitions*, in which *Heniz Scheidhauer* and *Werner Blech* were carried aloft in their *Horten Ho-3* all-wing sailplanes, and in the absence of a pressurized cockpit cabin suffered badly. *Blech* died, and *Scheidhauer* suffered severe frostbite. Well, at this time, our pilots in the *DFS* made their first flights in the upper heights over the Alps up to 9,500 meters [31,168 feet] altitude, and with the corresponding incontinence of the oxygen supply, temperature regulation, and so on, because sometimes they came down with frozen feet and hands, then came the development of the *DFS 54*. This had been about 75 to 80% finished when we were shifted over to the *DFS 228*. This was because in 1943, we didn't have any time anymore for pure civil soaring, and therefore, we set the *DFS 54* aside for more friendly days. I then devoted my full time to the development of the *DFS 228*.

MYHRA - So the *DFS 54* was set aside due to the demands of the war?

KRACHT - There was not an immediate need, because we had no more time for soaring and glider flying/sport. Soaring was considered a sport, and not a research program related to wartime.

MYHRA - How much did the *DFS 54* program come about through the problems experienced by *Scheidhauer* and *Blech* at the 1938 *Rhön/Wasserkuppe Summer Sailplane Competitions*?

KRACHT - *Scheidhauer* and *Blech* were friends and colleagues. So, yes, the *DFS 54* was prompted by their suffering at the 1938 *Rhön/Wasserkuppe Summer Competitions*. To help avoid the problems suffered by these two men, the *DFS 54* was designed with a pressure cabin, and from the work on the *54*, I had collected all information which was related to aeromedicine, in particular at higher altitudes, from a medical research doctor by the name of *Siegfried Ruff*. As a teenager, *Dr.-Med. Ruff* had been one of us and was a member of the Bonn sailplane group. He was a life long friend of *Reimar Horten*, too. Certainly, there were some errors in our design, but we found them out ourselves later on and corrected them. From the medical statements we had we learned that a man with training could go up to 8,000 to 9,000 meters [26,246 to 29,527 feet] without oxygen. Testing proved this later in the Himalayan Mountains while climbing the highest mountains without oxygen. But there are physiological limitations. At an altitude of roughly 20,000 to 21,000 meters [65,616 to 68,897 feet], pilots were exposed to the exterior pressure. Your blood, theoretically, would boil, and you would die. In fact, this limit is not computed in comparison to other pressures to the exterior. This is because the composition of blood delayed it up to 25,000 meters [82,020 feet]. But all this was known and was integrated first into our concepts at DFS, especially into the cabin of the *DFS 54*. With this knowledge, and then with the requirement for higher ability to sustain high "G" loads, we had gone to the concept of the *DFS 228* with the prone position pilot, so this was a continuous development beginning with the *DFS 54*.

MYHRA - What was the requirement for the high "G" loads?

KRACHT - The concept was that we would fly the *DFS 228* at the limit of its performance given by the thrust of its bi-fuel liquid rocket, and then the force in "G" would be high. After the end of the thrust you were going at high speed, and 4 to 5 "G's" were required to go immediately into the climb. This was the limit we thought about, because the aircraft at this time had a climb angle of 60° to 65°; you had to go immediately into a climb, otherwise you would remain too level and therefore unable to go through the *Mach* barrier. So, this was a clear requirement to withstand high "G" loads. Thus, we put all the knowledge which we had obtained from *Dr.-Med. Ruff* into the design of the cabin, and then with the cabin of the *DFS 228* we also built into it the "G" loads which it had to withstand. So, this was quite a continuous program of development. But also we had other projects, such as the "father and son." In parallel to all this came our cooperation with *Messerschmitt AG* on their *Me 328*.

MYHRA - On the other need a pressurized cabin in a high altitude sailplane, would people come to you and say that we have a need for this and that? Would they then ask *DFS* to design this or that, or would they bring a design to you and have *DFS* build it?

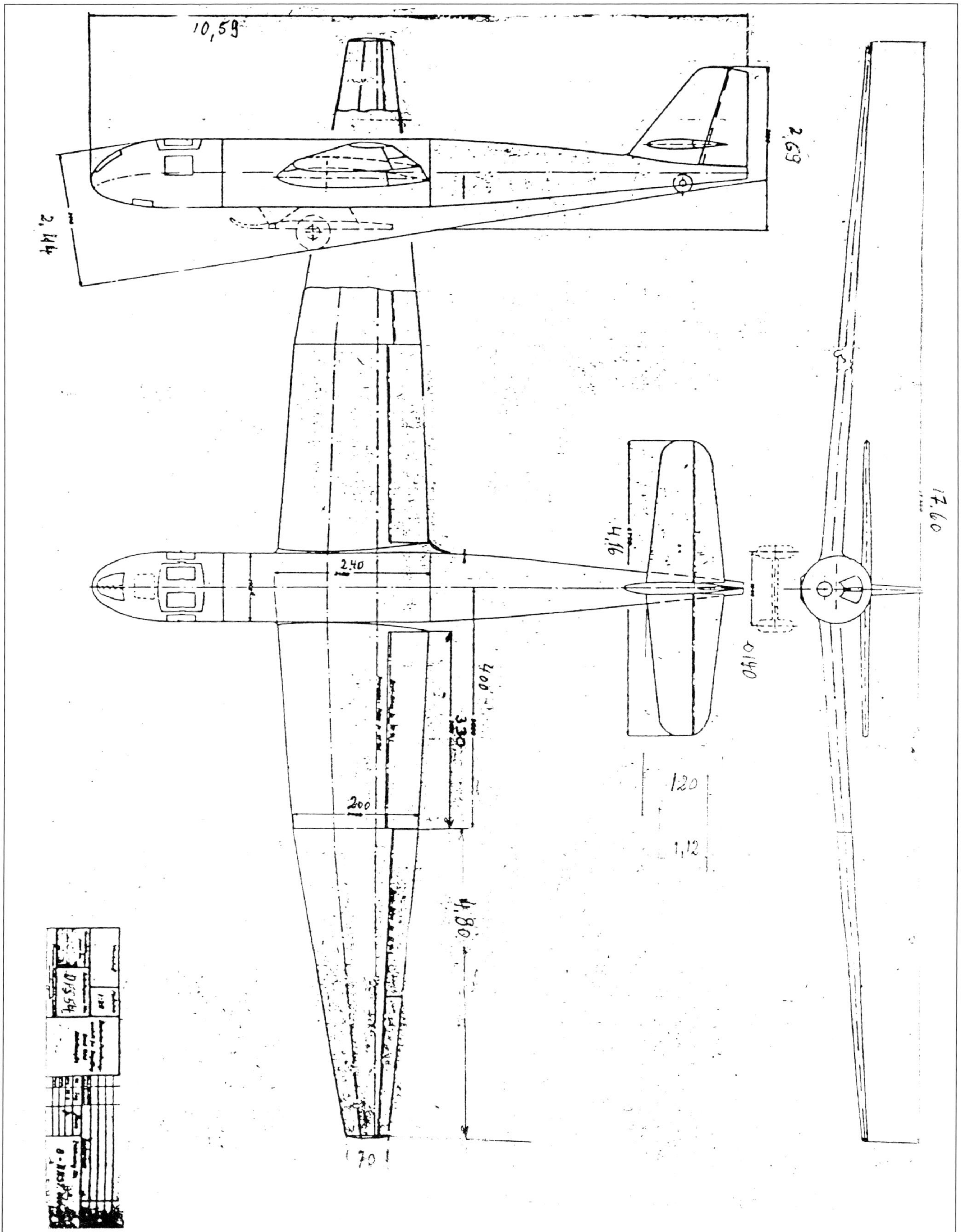
10,59
2,69
2,44
17,60
4,16
2,40
400
330
200
4,80
70
120
1,12

KRACHT - No. The development and need for any item came first from our soaring department at *DFS*. We were sailplane people right down to the core of our being. Our community had already been soaring at 9,000 meters [29,527 feet], and we were convinced that this was not the limit. But I had experienced that when they reached these heights and intended to go down, they could not, because they had put the aircraft at the highest level where normally they would dive at 6 to 7 meters [19.7 to 23 feet] per second. They were always flying level or climbing. So they came back to us and asked: "What are the physical limits of what we can do with our particular equipment, and how can we protect ourselves if we get beyond this barrier?"

MYHRA - Well, how would this type of safety at ultra high soaring research then be budgeted? I would imagine that *DFS* itself did not have the need for such research beyond the sailplane community. When your brothers in the sailplane community came to you at *DFS* regarding the need for a pressurized cabin, where would the funds come from to build it?

KRACHT - Very simple. *DFS* was an *RLM* institute, and was getting 100% of its funding from the *RLM*. Later, what they did with these funds was based on what *DFS* officials thought was in the best interests of the aviation community. *DFS* was relatively free to do what they thought was in the best interests of high altitude flight. Each year we received a certain number of *Reichmarks*. This meant that our people would be paid and equipment and supplies could be bought, and then we virtually did what we wanted with those funds. It was *DFS'* own priority. It was our business. I do recall that the *RLM* was not all that generous in terms of Reichmarks, but we got by. In the 1930s, when *Hitler* came to power, the first task was to have stability in prices and zero inflation. You can say that at the beginning they had the idea that 1,000 *Reichmarks* per month was a top level person. Later on these levels were slightly increased. On the other hand, pilots, at this time, received no risk payments for test flying. I had made numerous flight tests, and the total premium was 500 Reichmarks. That's all.

MYHRA - Since the *DFS* was funded through the *RLM*, did they have any priority in saying what *DFS* should work on?

KRACHT - Yes, but DFS tried to work out their own programs and present their own priorities to the *RLM*. The *RLM* approved them, or criticized them, or said that this task or goal was not possible. For example, I had some of my own personal ideas, and then I would discuss them with the *RLM* and see what I could do. But normally, the projects and priorities set in motion were made pretty much by *DFS* itself.

Opposite: A line drawing of the *DFS 54*, the origin of the detachable cockpit used on both the *228* and the *346*.

MYHRA - So there was also some liberty within *DFS* which allowed you to be creative and pursue some of your own sailplane ideas and interests? What percent was the *Luftwaffe* and what percent was *DFS*?

KRACHT - Yes, we had liberty. Prior to the start of the war in 1939, *DFS* had very few initial requirements from the *Luftwaffe*. With the coming of the war the *Luftwaffe* had some important requirements, and as the war became larger, increasing in scope, and increasingly dismal, too, then their requirements to use *DFS'* knowledge and personnel increased heavily.

MYHRA - What kind of relation did *DFS* have with the German aviation industry? Was it a close relationship?

KRACHT - Yes, it was a close relationship. By example, I can give you two cases. *Messerschmitt AG* was interested in testing the possibilities for a ground-attack aircraft powered by pulse-jets. It was known as the *Messerschmitt Me 328*, and its engines were the *Schmidt-Argus 014* pulse jets. Then *Willy Messerschmitt* came to us and asked us if *DFS* would cooperate with his company. It appeared that *Messerschmitt AG* wanted to build the initial project aircraft. It would have about a 6 meter [19.7 feet] wing span, which would be built out of wood, and then *Willy Messerschmitt* said that he would take a fuselage from a *Bf 109* and modify it to fit. *DFS* concluded that this would be too expensive for just a test aircraft. We felt that this proposed aircraft should have a circular fuselage cross section, and so we defined (drafted) the project and how it could be executed as quick as possible. Then we decided it could be developed out of wood, and the fuselage frame should be assembled out of steel tubes and covered with plywood. This had been done before. For example, the horizontal tail and the trim articulater came directly from the *Bf 109* and were delivered to *DFS* by *Messerschmitt AG*. However, with the "father and son" project there was a close association with *Junkers*. The idea for the "father and son" came initially from *Junkers*, and when the *RLM* said "don't spend money for the development of the combination aircraft, but go instead to *DFS*-Ainring, and there are people there who know how to do that." Then *Junkers* delivered a new *Ju 88* to us and *Messerschmitt* gave us a new *Bf 109*. *DFS* modified both aircraft to make the "father and son" possible.

MYHRA - Would either *Messerschmitt AG* or *Junkers* give you money to carry on this research for them?

KRACHT - No. There was no competition between them on this project. *DFS* merely did the job as instructed by the *RLM*. After a time we delivered the final product to *Junkers*. Our job was finished, and later *Junkers* was producing the modified *Ju 88s* for the "father and son" combination. There was no exchange of money between the two aviation companies and *DFS*. However, for us to do it required an order from the *RLM*.

MYHRA - All your projects with the aviation industry had to be approved by the *RLM*. Could any aviation company come to *DFS*?

KRACHT - Oh, sure. Right after we had developed the first coupling for towed aircraft, a man from *Junkers* who also had the same idea had approached the *RLM* about it, and he was immediately put in contact with us through the *RLM*. His interest in coupling also helped us at *DFS* and resulted in some increased interest by the *RLM*. I had made two patents with this coupling device to avoid trouble if *Junkers* should later oppose it.

MYHRA - Do you recall any projects that were not accepted by *DFS*?

KRACHT - No, not at this time. Because finally, the *RLM* knew the occupation of the different companies and what they could or could not do, and how fully they were employed. When those of us at *DFS* were fully employed, the *RLM* would send no more people around to us. So, once we were fully occupied we received no more orders from the *RLM*. We never once had to complain about a lack of work.

MYHRA - Were there many projects that *DFS* turned down for one reason or another?

KRACHT - Yes, but I don't recall any examples. By 1944, there were a lot of crazy ideas going around. For example, there was a proposal by a man who had some contacts in the *RLM* and who had insisted on developing a military glider with tandem wings. This I refused to do. This was the only idea that I refused to do which comes to mind. This project in particular had been submitted by a dentist who considered himself a great aircraft designer. Those of us in the *DFS* only considered him a dentist. But you have such crazy people in every place in the world.

MYHRA - There were people, too, who looked at the *Horten* brothers and their all-wing aircraft as crazy ideas.

KRACHT - The all-wing ideas of the *Hortens* had the possibility of flying. We know that they flew and flew well. *Northrop in the* USA also developed the all-wing. Even his giant bomber, the *XYB-49*, flew, but it was canceled. This idea of the pure wing is tempting when you first begin, but I'll give you another example why these are crazy ideas. Remember, you can make almost anything fly, whatever it is, but most of the solutions which are possible are turned down at the moment that you consider economic solutions to the problem. By example, in military aircraft they use the so-called "area rule" that means that the fuselage looks like the *Coke* bottle. Here in Germany we had some big aerodynamicists who proclaimed that the area rule should be applied to civil transport aircraft. From a pure performance point of view one might gain a few more km/h [mph] at extremely high speeds, but if you considered a civil airplane, such an aircraft must have a center compartment and a cargo compartment. The percent of the center compartment and cargo compartment is our common business with the consumer. Aerodynamics at this stage had nothing to do with it. Airlines want handling, freedom of installation of seats and for arranging seats in a given room, and so on, so the airlines will reject anything which does not give them a circular fuselage or a cylinder-shaped fuselage. So, aerodynamicists have limitations where they can locate the area-rule in the fuselage and where they will not be appropriate for civil aircraft as such.

MYHRA - Let me ask you a question about the *Horten brothers*. The all-wing turbojet-powered fighter, the *Horten Ho 9V2*, proved without a doubt that it could fly, but many aerodynamicists question that it could fly well enough to be an effective fighter. The reason being poor directional stability, a poor gun platform due to its lack of stability, and so on. Yet it was accepted by the *RLM* and put into serial production at *Gotha* as the *Ho 229*. The *Luftwaffe Quartermaster* was planning to outfit *JG400* with these machines. How could all this happen if the *Horten* all-wing had such serious faults?

KRACHT - You see, during its time period, there were some interests backed by strong supporters in crazy aviation ideas. All this shows is that the *Hortens* may have been very clever in maintaining personal relations with some important people in the *RLM*. So, many doors were opened to the *Hortens* which normally would not have been accessible to "normal" people. In any society you will have a certain number of, say, "playboys," and the *Hortens* with their all-wing designs were a sort of "playboys." You can, say, fly with the door off your barn. You can make almost anything fly. But you must consider how to do this, what are your needs, and requirements in terms of service. But to consider a fighter aircraft, well, this is a real consideration, because of stability needed for accurate shooting, especially when using big ammunition. The same stability is required when launching air to air guided missiles, and the flight behavior of the aircraft itself is critical. Without a stable base one should not even start a design.

MYHRA - The *Ho 9V2* went from an experimental prototype directly into serial production.

KRACHT - I know, but I do not have sufficient information on how the *Horten* brothers were able to do all this. *Walter Horten* was the clever one of this team. To me, this type of aircraft development, like the *Ho 9 V2* going directly into the *Ho 229*, is movement sideways and not forward. The *Hortens*, and the dentist I mentioned earlier, well, they are similar stories. I suppose that early on in the *Hortens'* career what they were doing was of considerable interest to people. In my mind it was planform which might work for particular missions and not necessarily for military, scientific, or civil, but which would not necessarily have any future in aviation. They appeared to derive considerable pleasure form doing this sort of work, but after the war the aviation community did not come back to the all-wing concept, and they still have not done so for over 40 years.

MYHRA - *Reimar Horten* could not find any work after *WWII* either in England, France, USA, and so on. He finally joined up with *Kurt Tank* in Argentina in 1946. *Tank* had gone to Argentina at the request of *Juan Perón* and to put into production *Hans Multhopp's* swept-wing fighter project of *WWII* called the *Focke-Wulf Fw Ta 183.* In Argentina it was called the *Pulqui II.* It looks very similar to the *MiG 15* of the USSR. What kind of man was *Georgii*, and what kind of mind did he have?

KRACHT - He was not a great thinker or innovator. He wanted to go to Argentina, too, and be with *Tank.* He was very worried that the French would not let him go. I remember one evening when I took him to Orley Airport in Paris to leave France for Argentina. He was also very afraid that the Russians would catch him. So, going to Argentina was really the right thing for him for a very simple reason: he found there his second wife, and they were very happy together, and this was the best thing that could happen to him. He later returned to Germany. He was a nice fellow, a supportive type of person.

MYHRA - What was Georgii's management style?

KRACHT - I would say that it was very easy working with him, because as long it was rocket propelled *Georgii* didn't interfere with your work. In fact, you have to remember that this was already late 1944, and communications had become quite difficult, so one normally thought it over twice before you tried to make a telephone call or even to make trip via automobile.

MYHRA - Any thoughts on how the test results would have been transferred back to the *RLM* or aviation industry on the *DFS 228?*

KRACHT - This was very clear: the test results would have been handled like any other test results in Germany in the form of write ups. This is the way we did that sort of thing then. The *RLM* decided on who should receive this information. By example, here is a copy of one of the very early reports on the air intake of a turbojet engine relative to the leading edge of the wing. Reports like this one were released quite regularly, and they were generally listed as classified, secret, or top secret. Persons/people with a need to know this research would be provided with a copy. This included people in the aviation industry and research centers. Normally, each copy was numbered, so people knew how many copies were outstanding and who had received it.

With the bi-fuel liquid rocket engine in the *DFS 228*, there was a problem in terms of the capacity of the fuselage for the storage of fuel. Also, it was hard enough with one *HWK* bi-fuel liquid rocket engine and difficult with two, let alone with four of them. This is because with the use of H_20_2...I don't know if you've heard of that, but it was the oxidizing of water and not of air. The main oxygen supply was a liquid H_20_2, which means concentrated hydrogen peroxide of 99% purity. All this stuff is highly explosive. I tried an experiment one day. It had rained, and it was really dark, and a teacup of that stuff placed on the grass and boom!

MYHRA - You mentioned the *DFS 332* was an idea for a flying wind tunnel?

KRACHT - The *DFS 332* was an aircraft with interchangeable wing sections, and the speed would be kept by use of an *HWK* cold rocket engine. This project had been built by my later chief in France. It was finished to about 80% competition by May 1945, so it never flew, and after the last half of 1944, I did not put any energy to promote it or even to make it fly. I knew in mid 1944 that with the war going the way it was going that this aircraft would not make any difference if it was built or not. The center wing had a depth of 5 meters [16 feet 4 inches] and had a variable thickness and profile. The *DFS 332* was a pure research aircraft. There were two of them, and they were being supported by *Jennissen* from the *RLM's LC-1.*

MYHRA - What type of carrier, or "mother" aircraft, were you thinking using for launching the *DFS 228?*

KRACHT - We would have continued using the *Dornier Do 217* aircraft.

MYHRA - Had there been any thought about using a *Heinkel He 177?*

KRACHT - No, because the *He 177* was at this time certainly Germany's most advanced aircraft flying—from every side. It had one weak point, a real essential weak point. They had these *DB 613s*, two *DB 603* coupled engines on the biggest propeller existing at this time. The installation led to overheating of these engines, and therefore the *He 177* was insufficiently reliable for us. The French, later on post war, had a variation of the *He 177* known as the *He 277*, and it was a clean four engine aircraft. But we at *DFS*, well, did not want to use a *He 177* at all.

MYHRA - The DFS 228's cockpit held a great many measuring devices. This is all pretty high-tech stuff for those days.

KRACHT - Yes, it certainly was. For example, all those different measuring instruments needed for the DFS 228 and 346, well, one could not just go out and purchase this equipment. We designed it and made it, too, ourselves. For example, look at these quick disconnect connections. We constructed the framework for the cockpit from wood using a nail-less process. Later on when we tested the cockpit we found that it was an excellent thermometer; that is, the pressure was variably proportional to the temperature. There was no loss whatsoever. The cabin was absolutely tight. We had a lot of very good people working on it, especially on the rear frame between the cockpit rear wall and the first forward balkhead on the fuselage. The flight controls were going through the rear frame.

MYHRA - Would there be a quick disconnect at that point?

KRACHT - No, no. You see here in this photo a lever which connects to these two half-moons, and then from the fuselage side came the other connection and pushed against it. So just friction held the two together. There were two levers side by side which were coupled. When you had to disconnect the cockpit there were no rigid connections, and it was easy to separate them.

MYHRA - Had this type of detachable cockpit ever been done before?

KRACHT - At that time? No, not that I know of. It was the first anywhere in the world.

MYHRA - Where did all these very original ideas come from?

KRACHT - From some very highly creative people. Take, for example, the connections for releasing the *Bf 109* from the *Dornier*. We had had explosive bolts, and we developed our own. The explosive bolt system was developed by an engineer named *Karl Bütter* from *Heinkel AG*. We used the explosive bolts, but we had problems with them. The difficulty came with flight testing. I had never lost a test pilot, nor we ever have a need to go to the hospital as a result of our test flight program at *DFS*-Ainring.

MYHRA - What was the procedure for separating the *Bf 109?*

KRACHT - There were three release struts. First, the pilot exploded the bottom bolts in the front strut, which then permitted the Bf 109 to ease up into the windstream. The incidence of the aircraft was then increased, and when the pilot had achieved a 30% to 40% incidence then the two rear bolts exploded, too. But with only the bottom bolt blown in the front strut, and as long as the incidence was not achieved, then the other two struts would remain fixed and would not explode.

MYHRA - So, when this first, bottom bolt was fired, the *Bf 109* pilot had to increase the angle of incidence and then other bolts were released, too?

KRACHT - Yes, and when the angle of incidence was achieved, the pilot had sufficient lift to go up and away from the lower aircraft. It was just this simple.

MYHRA - Once this one bolt was fired, how much time elapsed before the *Bf 109* was completely disconnected...three, four, or five seconds?

KRACHT - No, no. The switch for the explosive bolt separation was located on the pilot's control column/stick, and when he pulled it back he changed the incidence between 30% or 40%. At that point he was immediately released from the lower aircraft. Overall it might take between 5 to 13 seconds to be completed. The reason for the 5 to 13 seconds length of time depended on the angle of incidence, because the pilot had to pull back on his control stick. The order to explode the lower front strut bolt was on the pilot's stick. Even if the pilot exploded that one bolt he could still wait before pulling back on the control stick to change the angle of incidence and complete the release/separation. The pilot, say, could release the *Bf 109* immediately, or wait for a minute or more. It depended on the pilot's situation and how fast he wanted to disconnect.

MYHRA - The special instruments in the *DFS 228* and *346* were designed and crafted by *DFS*?

KRACHT - Yes, DFS modified existing air speed, temperature, Mach meter, and so on—all standard instruments which we modified at DFS-Ainring. We had a small machine shop at Ainring, and so we didn't sub-contract this work to anyone. We made all this in our laboratories. We purchased very few items, mainly modifying existing instruments or making them ourselves.

MYHRA - Where was the stabilizing parachute attached on the *DFS 228?* Actually, you had designed a dual parachute system. A stabilizing parachute and then a second one for braking/slowing the cockpit's descent.

KRACHT - No, we had in fact three parachutes. First was the cockpit canopy's stabilizing parachute. Second was a braking parachute, which released the pilot's couch mechanism on which he was lying. The couch mechanism slid out of the cockpit after it had stabilized. The third was the pilot's own personal parachute.

MYHRA - If the pilot was unconscious, would the canopy release automatically at a pre-set altitude?

KRACHT - Yes, it was fully automatic. Here is a photo showing the coupling mechanism for the rigid towing experiment.

MYHRA - What is this little picture/painting on the side of the towed aircraft?

KRACHT - It was a little swordfish. This was because a swordfish has a long bill, and, well, our towed aircraft also had this long coupling mechanism in its nose. Here we have a photo showing the mechanical items for the "father and son" arrangement. A better system was to be used later on for the *Me 328*, where you had the main strut under the wing and then one point in front.

MYHRA - When it came time to land did the *Bf 109*, for example, land on its own gear?

KRACHT - Yes, but for the *DFS 228, DFS 346,* and *Me 328* we used a skid, but it was all the same principal.

MYHRA - I've read where the twin *Schmidt-Argus 014* pulse-jets on the *Me 328* caused so much vibration that the aircraft was difficult to control.

KRACHT - No, not exactly. It essentially made only noise. One day over Salzburg I caused a lot of confusion in the streets when flying the *Me 328* with its twin *Schmidt-Argus 014* pulse jets operating.
Look here at the photo of the *DFS 228 V3's* cockpit. You can see the triple window, which was braced in such a way and was about 8 mm thick because the differential pressure was relatively low because the pilot was on full oxygen. Here you see a photo of the form of the pilot's couch and its different sections. In front of that you have here this contact and then...around the whole surface was a small steel ring where you had a simple knot. This knot would be handled by the braking, not the stabilizing parachute, that is, parachute #2. Then the pilot also opened the window, and then a few seconds later at an altitude of 9,000 feet [2,743 meters] the couch was pushed out of the cabin by air pressure. Now the pilot, though still lying on his couch, was free of the aircraft. Then, moments after being pushed out of the cockpit he was released from the bed and automatically his own parachute opened. That was parachute number three.

MYHRA - So if the pilot was unconscious he'd still be saved through this automatic sequence?

KRACHT - Yes, I was placed in charge of the *DFS 228* essentially due to my experience in aircraft towing. In addition, I was selected because I had a lot of experience in flight testing. I can tell you when I discussed this automatic ejection system with the U.S. Army CIC intelligence after the war and how we *DFS* engineers would like to continue this job in the USA, the CIC said they could not find a company willing to do this type of rocket propelled aircraft research in the USA.

KRACHT - The *DFS 228* was never designed or intended to go to Mach 1.0, but only designed to reach between 0.70 to 0.75 *Mach*. So, our *DFS 228*, even at *0.75 Mach* with its profile, wing structure, and overall aerodynamics, it was able to reach its intended design speed at an altitude of 100,000 feet.

MYHRA - Are there lifting eyes in the wing of the *DFS 228?*

KRACHT - Yes, but I don't have any photos which show them. These attachment points had already been designed and had to work. This was quite unusual at this time, no? These projects at *DFS* gave us engineers some very good experience. By the way, in the cockpit of the *DFS 228V1* was a little window in the nose through which the pilot could escape if he had to, but it would have been tough to get out.

MYHRA - I'm amazed by all the new engineering technology that went into this aircraft.

KRACHT - If you do something new without a lot of new things, then what have you gained?

MYHRA - When you build a one of a kind research aircraft like the *DFS 228*, and naturally you are concerned about pilot safety, how were you able to test all this new equipment/ideas to be sure they'd work as intended during flight?

KRACHT - Certainly, we did not have an organization in *DFS* which was timid in trying out new ideas and things. Today, that is not the case. You can see that this timid approach doesn't work, either (the British stopped their manned supersonic flights in favor of unmanned in 1946, because they feared for the pilot's life. Later they had to return to manned flights in order to collect the data they needed.) *DFS* was a very small organization. We only had about 22 or 23 people working on the *DFS 228* project, total, ever.

MYHRA - How many hours per day would you and your staff work?

KRACHT - At this time the normal work time was roughly between 15 and 17 hours per day and six days per week, or about 100 hours per week all together.

MYHRA - Here is a photo of an aircraft designed for prone-flying. It appears that it has an *RLM* designation number of "*108-54?*"

KRACHT - Yes, it was our *DFS 54*. Anyway, this aircraft was designed in the early Autumn of 1939 for high-altitude soaring. It would have had a pressurized cockpit made out of wood for flights above 11,000 meters [36,080 feet]. This was the limit beyond which sailplaners could not safely go without a pressurized cockpit. But due to the war it was never finished, although a mock-up had been 80% completed at war's end. The *RLM* established four aircraft design classifications, and for the one for powered aircraft it was "*8*." For sailplanes it was "*108.*" *This* is why in the photo you see the *RLM* classification number *DFS 108-54*.

MYHRA - How often would inspectors from *RLM's LC-1* come by to review the design work on the *DFS 228?*

KRACHT - Not often. In fact, on the *DFS 228*, we had no regular contact with any official from the *RLM*, but from time to time someone would come by and talk, and then we'd show them the *228's* design drawings. Then, too, I would go frequently to the *LC-1* and *LC-2* to confer. Plus, *DFS* held regular meetings with the *RLM*, and progress reports would be given at this time on all our work within the *DFS*. But at this time we did not have any strict controls over us by anyone in the *RLM* for the *DFS 228*. On this aircraft we were rather free! We had a task to do and we did it. The DFS 228 was intended for military purposes, but only as an ultra high altitude reconnaissance aircraft.

MYHRA - What type of reconnaissance role was intended for the *DFS 228* in late 1944? By that time all the fighting was within your own country, and you knew your own country.

KRACHT - Yes, but the *DFS 228* program had started already in 1942, so we continued working on it, all through the war right up to our country's surrender on 8 May 1945.

MYHRA - Yes, but what need was there for a high altitude reconnaissance machine like the *DFS 228* even in late 1944?

KRACHT - Well, it was still interesting to have such an aircraft like this even at this time of the war, and the *RLM's LC-1* decided to pursue the possibility. At first there was some talk of perhaps towing the *DFS 228* aircraft like a sailplane up to 40,000 feet [12 kilometers], releasing it, and then the aircraft would first go on under its own power to the limits of its intended *Mach* number of *0.75 Mach*. Then it could have reached an altitude of about 80,000 feet (24,384 meters). By then the *DFS 228* would be flying at an altitude between 40,000 and 80,000 feet [12 to 24 kilometers], and at that altitude there would be very little air, so we'd experience no enemy jet powered, and certainly no piston-powered fighters to interfere. At this altitude you knew you were safe. The *DFS 228* aircraft would be all alone. When you came down from this altitude you would relight the *HWK 509* rocket engine and make a second jump, a steep climb to achieve a good gliding angle and back up to maybe 95,000 feet [29 kilometers] this time. All this was absolutely possible with the *DFS 228*, and in the meantime you have covered maybe some 400 nautical miles [646 kilometers]. So, all this time you would have been out of reach of any fighters. Of course today, such a flight would be different.

I recall that *Tileneus* from the *RLM* was a very ingenious man, and he was involved in our high altitude research work at the *DFS*. He supported the *DFS 346*, too. There was another person, *Professor Dr.-Med. Siegfried Ruff.* He was a Doctor of Medicine, and it was alleged that he submitted prisoners of war to high-altitude testing experiments on the ground. But except for selected pilots which volunteered or had accepted this testing as part of their job, *Ruff* had done a lot of testing on himself. I worked with the "son" part of the "father and son" on explosive tests on the cockpit release mechanisms. That means exploding the cockpit bolts at high altitude and the release of the cockpit and the impact on the pilot(s). I can tell you that if you make a jump at 45,000 feet and float down to safety in only a few seconds then you'll survive.

MYHRA - How intense would the explosion be when the bolts were exploded to release the cockpit cabin from the fuselage? How loud and how much force?

KRACHT - No noise at all. Well, if you were in an aircraft with a running engine, it would be about the same noise as you have while flying. No additional sound at all.

MYHRA - When the cabin was released, the pilot still had his own oxygen supply?

KRACHT - Yes. Everything the pilot required for flight was all inside, so when the cockpit was released from the fuselage, well, nothing changed in so far as the pilot's life support systems. We used the standard cylindrical oxygen bottles and not the little round balls. The main oxygen generators were in the cockpit chamber. This was for the simple reason that the oxygen supply would have the same temperature as the interior.

MYHRA - All the cables, wires, lines, and so on going back to the fuselage came through the cockpit cabin, and they, too, had a quick disconnect ability?

KRACHT - Yes, all the cables, wires, lines, and so on came through the cockpit cabin wall. Yes, they also had a quick disconnect, which had been standardized throughout the aircraft industry at that time.

MYHRA - How had all these systems been tested for their work ability prior to installation in the *DFS 228?*

KRACHT - These quick disconnect systems were already standard in the industry. As far as I know, we did not even test them. We already knew how to go out with the cable with a slight twisting effort, not a pulling effort only, so it would break away very easily when needed. But the cables were very fixed and solid if you did not twist. We made a connection with a twist so that they would disconnect. We had a large number of bolts with the "father and son" combination, especially on the *Ju 88* and *Bf 109.* We also had some direct links and then simultaneously a connection for the lower engines on the *Ju 88.* Oh, by the way, we had been able to obtain 1,500 kilograms [3,307 pounds] of thrust from our *HWK 509A*, which was used in the *DFS 228*

MYHRA - The *DFS 228* featured a prone pilot flying position. What training aircraft was used to give the pilots experience in a prone flying position?

KRACHT - We had an aircraft which had been designed by a study group from the *Technical University*-Berlin called the *Berlin B-9.* This was a small, twin-engine aircraft with 200 horsepower each which was designed to test high "G" accelerations. So, it was a very robust aircraft, and you could dive and reach the highest possible speeds, then recover, and experience some 5 to 6 "Gs." This aircraft served as a test bed, and our pilots flew it often. My friend *Hans Niese* had come from the *Technical University*-Berlin and knew the people who developed the *B-9.* We flew that machine extensively. I had some doubts that a pilot would be able to make long-range flights in the *DFS 228* of 8 to 9 hours duration in a prone position. This clearly was a possibility, and we had tried it and the pilot was able to withstand considerably higher accelerations than a pilot that was seated conventionally.

MYHRA - What were some of the disadvantages of the prone flying position?

KRACHT - Fatigue in the neck due to the chin support for the head. Overall, it was quite comfortable if one's flight only had to last several hours.

MYHRA - Would this prone flying position have been changed later in favor of a conventional seating arrangement? Would it have been modified to a more conventional sitting position?

KRACHT - No. Clearly we came to the conclusion after a lot of testing that the prone position optimized performance, but the difficulty would be in flying long duration flights.

MYHRA - How many *DFS 228*s were under construction at *Siebel* between 1944-45? Also, how many were planned to be built?

KRACHT - We had some 25 *DFS 228s* on order. I recall that three or four *228* aircraft had already been delivered on this order. In the production process, after the definition of the aircraft, say from the detailed design, I don't know how many would have been constructed overall.

MYHRA - Once the *DFS 228's* design had been completed and turned over to *Siebel*, what did you do next?

KRACHT - I still had the "father and son" project, and a new project called the *DFS 338.* Normally, for flight testing the "father and son" at *DFS* we had a *Do 217 K3*, and later on for definite tests it was *M2.* Our *K3* was the normal and latest version of the *Do 217*, and the *"M"* series was very popular as a high altitude aircraft, in particular for reconnaissance flying and for over water, for which it had a longer wing span and special engines and so on. The same principal is used for ferrying the *Space Shuttle* back to Florida on the specially equipped *Boeing 747.* Once we gained experience this "father and son" arrangement became quite a normal operation.

This picture shows a profile of the side panels in the *DFS 228's* cockpit. You can see the supports for the front windows are fixed on both sides. With the system of fixing them, to work them in a closed position. Here is the main structure, and you see the points where it is fixed on the rear frame. So, from that wooden frame, it is really a monocoque structure.

These photos show the head or chin support. The pedals were operated by the pilot's feet, and here you see the main control stick. There are two of them: one on the right hand side, and the other on the left hand side. The sequence is something like this: the first pedal was the landing skid, the second one was the trim, and the third was air brakes. One, too, is for the landing flaps. Notice that each handle has a different shape so that the pilot can differentiate between them. Here is a photo of the instrument panel, and in front of the pilot is a rear view mirror. When the pilot looked into that mirror he'd find some things, and notice that the wording is inversed so it read correctly in the mirror. The pilot could normally turn around a bit, but it was planned that he shouldn't, and this was the purpose of the mirror.

Here is a photo of one of the DFS 228's electrical panels. Notice all the switches. The pilot would merely have to extend his arm in order to activate them. This panel was on the right hand side of the cockpit. On the left hand side there were other instruments. Here is a photo showing the underside panel. We had others, too, I think. This photo shows the lower side, and the other photo shows the upper side panel. Another photo here shows the electrical and instrument panel. Here is a photo showing some details from inside the cockpit structure itself. Shown, too, are the oxygen bottles—a combination of tanks and balls.

We had made a very advanced study to be sure that everything would fit. We had done this in test aircraft and also made test flights in a *Berlin B-9* with the pilot in the prone position. This photo shows the location of the parachute and the other equipment and the sequence of that stuff. In this flat structure that was between them was the parachute, so everything was all nicely located.

In this photo, the parachute is located up here, and you see it on the other side of the oxygen bottles. You see also that there are different parachutes and equipment already installed on both sides of the pilot. Access to the pilot's rack was through a nose opening. The nose window was opened, and the pilot could enter and egress. This photo shows some structure, and you see here the panel, and there are the parachutes. Here is a cylinder for pressure air, which would be opened to put air pressure in the cabin. Below, you see a cable that unwound, allowing the pilot and his bed to be let out by cable. Inside here you see in the photo one of the levers for pulling the pilot out. The whole thing with the front window moved forward then out of the nose and away from the aircraft. During separation, the cockpit cabin would break away from the fuselage. In this photo is a circular link. We also saw the "O" ring which was placed in the opening separating the cabin from the fuselage. Here is a photo showing the system for rigid towing. That is the towing mechanism. Here you see my example of tight flight controls. This photo shows the frame as seen from the rear.

This photo shows how the pilot could rest his feet without getting fatigued. As the pilot was lying there his feet could hang down, and he felt better. There was a lot of human engineering involved. We went to great lengths to avoid pilot fatigue. This photo shows the explosive bolts. This photo shows where the explosive charge goes into the cylinder. This photo shows the clam-shell operation of the rigid towing connection system. In this picture is the chin rest of the pilot on the *DFS 228*. The chin rest was fully adjustable, and also there was some spring movement in there. So, this equipment was designed to be highly functional and relaxing. As a result, the whole idea was that you did not need to bring any effort into flying the *DFS 228*. Here you see a photo of the fuselage structure, the supporting structure, and the location of the different equipment. Here is a photo showing our system of foot pedals on the *DFS 228*. One of the levers we see in the photo was the landing flap, and the other was for the metal landing skid. Here is a photo of the apparatus for the towing aircraft. The receiving ball enters, and the mechanism clamps down on it.

This photo shows the nose thruster on the towed aircraft. Here was the main stick, or pilot's control stick. It was a very complicated arrangement. This picture shows more details of the pilot's prone seat. This photo shows the slide on which the pilot's bed moved in and out. It required much physical support, because it had to be strong if it were ejected. This photo shows the swordfish—because of its long bill we therefore called this aircraft with its long nose a "swordfish."

MYHRA - It appears that the *DFS 228's* profile is similar to that of your sailplane the "*Rheinland*." I can see the influence of the "*Rheinland*" in the *DFS 228.*

KRACHT - Yes, I think so. Here is a photo of our *Klemm* aircraft.

MYHRA - About how many flights did the *DFS 228* aircraft make?

***Rolf Mödel,* fabricator of the *228's* pressurized cockpit cabins. *DFS*-Ainring, July 1942.**

KRACHT - About 200 in all. Here is a series of photos showing the coupling process in mid-air. See the tow cable? This photo shows the coupling nearing completion and then fully coupled.

MYHRA - How much resistance could the coupling mechanism sustain?

KRACHT - About 4 to 5 tons. The cable was the weakest point of the whole system. It would break before anything else would. Fully extended, this proboscis would extend out about 1.5 meters [4 feet 9 inches]. There was an electric power wrench on the cable to help wind in back in quickly. Later on we used a *Junkers Ju 90* as a towing aircraft. *Hanna Reitsch* did not want to participate in these rigid towing experiments, but we showed these people anyway that it would work.

End of *Felix Kracht* Interview

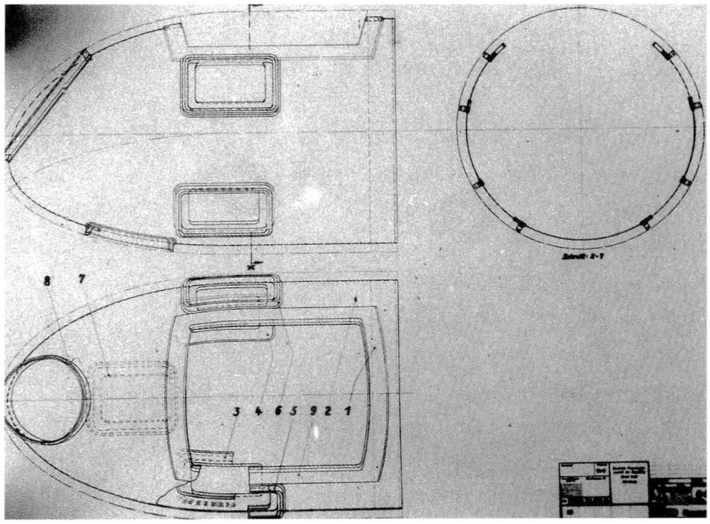

A pen and ink 3-view drawing from *DFS*-Ainring of their all wood detachable, pressurized cockpit cabin for the *DFS 228 V1*. Pilot entry and egress was accomplished through the large hinged dorsal hatch. In addition to the large round plexiglass port hole, the cabin had five other reviewing ports: one on the bottom of the cockpit and two on each cabin side wall.

Rolf MÖDEL
28 May 1988
Ingolstadt, West Germany

MYHRA - I've been told that you built several cockpit cabins for the *DFS 228*.

MÖDEL - Yes, based on the plans given to me by *Flugbaumeister Felix Kracht,* I built two versions of the cockpit cabin. Both versions were built out of wood. The first version, for the *DFS 228 V1*, had three round windows as shown in your photographs. The pilot sat up-right in the conventional manner. However, the pilots found this cabin most difficult...almost impossible to see much of anything, and for this reason we built a second model. The second cabin version was for the *DFS 228 V2*, and it, too, was built out of wood. It had only one large single-piece plexiglass window as shown here in your photographs. In this version for the *DFS 228 V2*, the pilot was placed on a couch. The large, single plexiglass window went a little aft of the pilot's head, so now he could see 180% horizontal, that is, from side to side. In addition, the pilot had about 120% or more vertical viewing, so the pilot now had almost a full view.

***Rolf Mödel*, Ingolstadt, West Germany. May 1988. Photo by the author.**

The *DFS 228* was designed for photo reconnaissance at great heights, up to 82,000 feet [25 kilometers]. At this height there would be no enemy aircraft to worry about. None. There would be no *Flak* danger, either, for at that time *Flak* shells reached a maximum altitude of about 32,000 to 39,000 feet [10 to 12 kilometers]. The *DFS 228* was being designed to be used only for photographic reconnaissance and nothing else. It had no armor or armament. The *228's* engine was a single *HWK 509*D bi-fuel liquid rocket engine. This machine was not designed to lift off the *DFS 228* itself, but was to be carried up to between 26,000 to 32,000 feet [8 to 10 kilometers] altitude on the back of a *Dornier Do 217K* twin-engine bomber. Then the *DFS 228* would be released and its *HWK 509D* ignited, and the aircraft would climb to about 82,000 feet [25 kilometers] altitude From this point on the *DFS 228* would enter its "glide phase" of between 62 to 93 mph [100 to 150 km/h] while dropping 3,200 to 6,500 feet [1 to 2 kilometers]. At this point the pilot would ignite the *HWK 509D* rocket engine again for a second time and start the "power up phase" again. The pilot would power the *DFS*

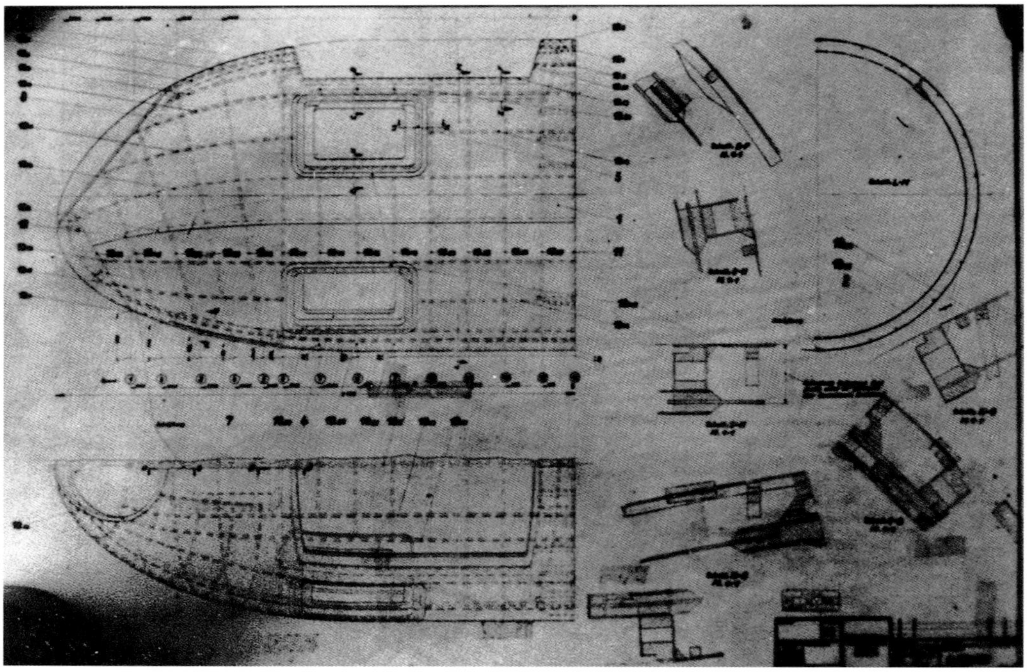

A poor quality pen and ink drawing from *DFS*-Ainring featuring the port side of the *DFS 228 V1's* cockpit cabin.

228 up to about 82,000 feet [25 kilometers], and this would take about 5 minutes. Then the *228's HWK 509D* rocket engine would be shut down, and the "gliding phase" would begin all over again. This system of flying, or flight approach, was called "*sagezahn*," or saw tooth. In fact, the little painting on the side of the *DFS 228's* fuselage showed a swordfish inspired by Eugen Sänger. This fish was our emblem for power up, sail down, power up, sail down, and so on. The number of times this flight approach could be carried was dependent on the quantity of rocket fuel carried on board.

MYHRA - What was to be the total flight distance (total range) of the *DFS 228* after being released from the *Do 217K* and its to return to base?

MÖDEL - I don't know for sure, no one really did, but the total range possible as anticipated might reach as much as 620 miles [1,000 kilometers] in total range.

MYHRA - Did a *DFS 228* prototype ever fly powered by its bi-fuel liquid rocket power unit?

MÖDEL - No. However, we had transferred one of our prototypes to *Siebel Flugzeug*-Halle. *Fritz Siebel* had a contract from the *RLM* to build additional *DFS 228* prototypes out of aluminum. I remember that we at *DFS*-Ainring built three *228* prototypes out of wood.

MYHRA - I did not know that a *DFS 228* was sent to *Siebel*-Halle for serial construction. So, construction plans for both the *DFS 228* and the *DFS 346* went to *Siebel*?

The *DFS 228 V1* in a landing approach and featuring the port side of its cockpit cabin. The *V1* carried a painted on symbol of a swordfish suggested by Eugen Sänger. Its teeth are in the form of triangles and form saw teeth like the *DFS'* power up glide down, power up glide down.

A poor quality photograph of the *DFS 228 V1's* starboard side cockpit cabin post war. It is laying pretty much broken on the ground in a RAE-Farnborough, England, junk yard.

MÖDEL - Yes, both aircraft types went to *Siebel*-Halle, and I don't know exactly why this came to be. We built the initial *DFS 228* prototype at Hörsching. However, *Siebel*-Halle was to build numerous other *DFS 228s*, and out of aluminum, too.

MYHRA - How many *DFS 228s* was *Siebel*-Halle asked to build...5, 10, 25?

MÖDEL - I don't know the exact number anymore. Construction in wood is much different than construction in metal, because metal requires highly skilled workers. Metal construction generally takes longer, too. At *DFS*-Hörsching we were in the process of building five examples of the *DFS 228* out of wood. Three had already been completed by the end of the war. I do recall that in our first proposal to the *RLM*, which came to be known as the *DFS 228 V1*, we proposed that the aircraft should have a cockpit with the pilot in a standard, conventional siting position. The idea to place the pilot on a couch in the *DFS 228 V2* came from research work done during 1937-1938, at the *Technical High School*-Stuttgart, known as the *Flugtechnische Fachgruppe*, or *FS*. The aircraft they built, designed by *Felix Kracht* and *Hans Niese*—they were members of that *FS* by the way—was a sailplane known as the *FS-17*. In this aircraft, the pilot laid on a couch. The *FS-17* machine by *Kracht* and *Niese* should not be confused with the twin-motor prone flying *Berlin B-9* from the *Technical High School*-Berlin. Even as far back as 1933-1934, there had been experiments with a pilot flying a sailplane while in a reclining (siting) position and flown by *Karl Bauer* (later, in the mid 1940s, chief of test pilots at *Messerschmitt AG*, Augsburg). This sailplane was known as the "*Fledermaus*," or field mouse. In addition, we at *DFS* thought that due to the high altitude and different maneuvers required, pilot "black out" was a possibility in the *DFS 228* and *DFS 346*. So, we believed that a prone flying position would be better than a siting position. We found that the prone flying position was good, and that with a chin-support the pilot did not suffer too great a fatigue.

MYHRA - Would not the pilot's neck become sore after a few hours flight-time from the chin- support?

MÖDEL - No. With the second cockpit cabin, for the *DFS 228V2*, the pilot now had very good visibility. However, the *V2's* cabin became a perfect circle-section, and its construction was very difficult, especially to make the whole cabin rigid and tight with wood. But the circular section was best for pressurization. The whole cabin framework was constructed out of wood. Its outer skin was plywood. We glued the plywood on to its wooden circular framework. All in all, we were seeking the lightest minimum weight possible.

A direct port side view of the *DFS 228 V1* with its hydraulically articulated metal landing skid in the full down position.

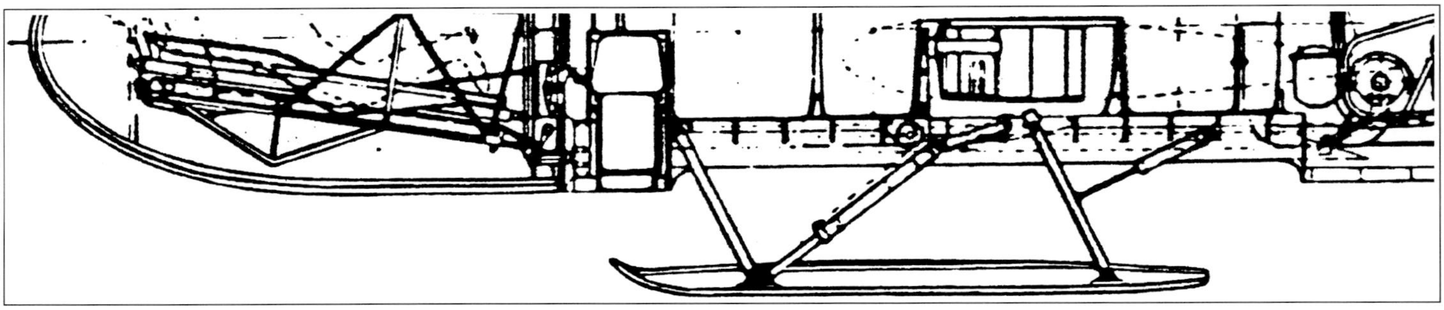

A pen and ink 3-view drawing by *DFS*-Ainring of the *228 V1's* hydraulically articulated metal landing skid. Notice that when it retracts up into the fuselage bottom it goes into its own little box, thus making for a flush fit with the surrounding fuselage.

MYHRA - About how much did the cabin weigh?

MÖDEL - I think about 174 pounds [79 kilograms] complete. I recall that when we had finished the first *V2* prototype cabin out of wood we took it outdoors and proceeded to pressurize it until it blew up. It could take up to 2.0 atmospheres of pressure before exploding. The exact shape of the plexiglass nose in the final version, used on the *DFS 228 V3*, was longer than shown here in your photos. Even its construction was a little different from the first two models.

Myhra - Did this final cabin version for the *DFS 228 V3* come to appear more like the cockpit cabin used on the *DFS 346?*

Mödel - Yes, exactly. With the *228 V3's* cabin, there was an outer nose of glass, if you will. It was a one piece window, and it was longer, more streamlined, and formed from plexiglass, and it was placed over the cockpit's inner plexiglass cone. It also helped cut down on the inner plexiglass fogging over, as well as icing over. So, we had one outer streamlined window, and behind that the pressure window for the cockpit cabin. I don't remember exactly if the space between the outer and inner cabin windows was pressurized to the same pressure as that in the cabin. It may have been. Nevertheless, I recall that this outer plexiglass nose was very thick in order to withstand the air pressure against it at high speeds. It was not difficult for us to make that one-piece plexiglass window, either.

MYHRA - Did either the *DFS 228* or the *DFS 346* have that long nose probe fitted to the cockpit cabin that we sometimes see in photographs of the *DFS 346* in the USSR?

MÖDEL - No, we did not use such a device on any of our prototype cockpit cabins for our *DFS 228.* That probe was probably added by our colleagues who were taken to the Soviet Union post-war.

The *DFS 228 V1* about to set down on the grass with its landing skid fully extended. The aircraft appears to be painted in a light gray stain. Notice, too, its civil registration has not yet been applied.

The *DFS 228* painted a cream color which was common to sailplanes of this era. Notice, also, that civil registration has not yet been applied to its wings and fuselage, however, its swordfish emblem by Sänger had been painted on the cockpit cabin's side wall.

MYHRA - Was it difficult for the pilot to enter the 3rd prototype cabin with its couch and then make himself ready for flight?

MÖDEL - Yes and no. However, the pilot had to have help or assistance getting into the cabin and positioned down on the couch. Then he required further assistance in getting himself hooked up to the safety harness, earphones, oxygen systems, and so on. Finally, the front plexiglass window was made ready to close. It was designed so that the pilot could turn a handle, and the window was seated, sealed, and made tight. It was a bit complicated, and he had to all of this with only one hand.

MYHRA - I've heard stories that the seal around the window leaked, and that it continued to be a problem with the *DFS 228.* Was this one of the reasons why the aircraft was not flown with its bi-fuel liquid *HWK 509D* rocket engine?

MÖDEL - Oh yes, perhaps the biggest reason. We found it to be very difficult to make the window seal tight enough to maintain the required cabin pressurization. The seating and sealing process got to be very complicated, so much so that it became nearly impossible for us to correct it prior to war's end.

MYHRA - In what year was the first cockpit cabin built?

MÖDEL - For the first version of the *DFS 228,* the *V1,* it was between 1942 and 1943. Many people told us that it would be impossible for us to make this cockpit cabin's front window tight enough and also release from the fuselage should the pilot need to escape at a high altitude. Perhaps. But we doubted that the pilot would experience any difficulties in exiting the cockpit during flight.

MYHRA - Was the pilot escape procedure ever tested or practiced during flight?

MÖDEL - Yes it was, by the way. Also, our *DFS 228* aircraft project was top secret.

MYHRA - Who in the *RLM* was the chief on the *DFS 228?* Was it *Oberst Siegfried Knemeyer*?

MÖDEL - I don't remember. *Oberst Knemeyer*, by the way, had been at the *Akaflieg*-Berlin. He was very bright, quick witted, and a very skilled pilot.

MYHRA - Was the cockpit cabin for the *DFS 228V3* the same diameter and length as the one for the *DFS 346?*

MÖDEL - I believe that they were the same. Really, the fuselage on the *DFS 346* was about same as on the *DFS 228.* About the only difference between the two fuselages was the *HWK 509D* engine's

A beautiful photo of the entire underside of the *DFS 228 V1* painted in a cream color. Notice the lower port and starboard plexiglass windows in the cockpit cabin. The black square near the nose is a viewing port allowing the pilot to see the approach ground when making his landing approach.

A poor quality photograph of the *DFS 228 V1* fully configured, including its civil registration of *D-IBFQ*. Seen here as it is preparing to land on *DFS*-Ainring's grass air field.

location in the *DFS 346*. Otherwise, both aircraft cabins and fuselages were pretty much the same diameter and length. I believe that we even used the same drawings for both aircraft for the cabin and fuselage.

MYHRA - So, the *DFS 228* and *DFS 346* fuselages were similar? That is, the same diameter, length, wing root position, and so on?

MÖDEL - Yes, exactly the same size physically. On the *DFS 346* the rear portion of the fuselage had been changed, enlarged really, to accommodate the newer *HWK 509* and two dual combustion chambers. Actually, its two thrust nozzles were of the same diameter. Also, the vertical rudder and horizontal tail-plane were changed somewhat. The *DFS 346* carried a greater volume of *C-Stoff* and *T-Stoff* than did the *DFS 228.* But this was just an internal rearrangement. So, this was about all of the differences between the two aircraft. You can see this in your photograph showing the damaged *DFS 228 V1*—the *HWK 509D* had been customized for installation. Normally, the combustion chamber (nozzle) was connected by a long thrust shaft to the mixing chamber. But in the *DFS 228* we separated the combustion chamber from its mixing chamber. There was no thrust-tube/shaft in the *DFS 228.* In the *DFS 346*, the two combustion chambers were mounted in the same fashion...one above the other, and no thrust tube. Fuel lines and pipes were also lengthened to reach the mixing chamber, which was mounted about aft of midship, just as it was with the *DFS 228.*

MYHRA - In this photo of the engine area of the *DFS 228*, it looks as if the combustion chamber has had some support rods welded on to it, and that these rod ends were attached to a bulkhead. I don't

The *DFS* created the "*Mistel,*" or father and son carrying arrangement. On top of the *Do 217 K3* is the *DFS 228 V1.* It appears to be painted in its cream color without civil registration or national markings. This starboard side nose view was taken at *DFS*-Ainring. Notice that the swordfish emblem does not appear on the *V1's* starboard side.

Notice that the *Sänger* suggested *DFS 228 V1's* swordfish emblem appears on the aircraft port side fuselage nose and not on the starboard side.

see a thrust tube, and you can see, too, the fuel lines/pipes run forward to the mixing chamber.

MÖDEL - Yes. I recall that the mixing chamber was located near the center of the aircraft's fuselage.

MYHRA - Some people say that there were two separate *HWK 509D* bi-fuel liquid rocket engines in the *DFS 346* and that they were mounted one on top of the other. Was this the way it was?

MÖDEL - No, but there were two separate nozzles. The *DFS 346* didn't have the *HWK 509C* bi-fuel rocket engine, either, but its dual chambers were of the same size, same diameter. May be early in its development, the *DFS 346* might have been planned with two complete *HWK 509B-1* rocket engines. But the *DFS 346's* final plans did not include two separate engines...only one, and I believe that it was custom made specifically for the *346's* high speed needs. Many details are changed, which is a normal process in aircraft development. Therefore, drawings cannot be completely relied upon. These drawings showing the *DFS 346* with two complete *HWK 509B-1*s one mounted above the other is not correct.

MYHRA - Where in my photos of the *DFS 228's* cockpit are the flight instruments located?

MÖDEL - Well, they are not really visible in your photos, and they were not placed in one location as they normally are in most aircraft. The necessary flight instruments were located under the pilot's couch and just below and to the rear of the chin rest. There was a mirror to help the pilot see these instruments, so all the pilot had to do was look down into the mirror and he'd see his flight instruments in the mirror.

MYHRA - What are all these switches for?

MÖDEL - They were electric starting and running switches and associated with the *HWK 509D* bi-fuel liquid rocket engine.

MYHRA - What are these bottles used for?

A view of the *DFS 228 V1* atop its *Do 217 K3* carrier. The *V1* piloting position was the standard upright seating. Notice the individual standing in the open hatch of the cockpit. The hatch was hinged, or it could also be lifted up and removed.

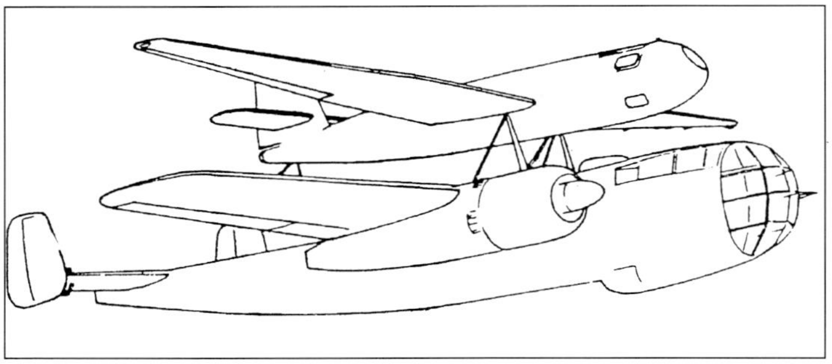

A pen and ink drawing from *DFS*-Ainring featuring the *228 V1's* starboard side as it would appear high atop its *Do 217 K3* carrier aircraft.

MÖDEL - For pilot's oxygen and air to pressurize the cabin.

MYHRA - What are these instruments which are located on this vertical panel?

MÖDEL - I don't remember exactly, but they were of secondary importance to the pilot. He would view them also in a mirror. This upper box over the back of the pilot held his personal parachute.

MYHRA - Where was the cockpit cabin's stabilizing parachute located?

MÖDEL - You can't see it in these photographs, but it was attached to the aft portion of the cabin.

MYHRA - What were these controls in this photo of the pilot's couch?

MÖDEL - The hand wheel you see was for adjusting the trim. The stick was the "control stick" for elevation, and so on. The pilot had foot pedals which operated the rudder control mechanism.

MYHRA - Where were the controls for the *HWK 509D* bi-fuel rocket engine?

MÖDEL - I don't remember exactly where they were located.

MYHRA - Do you have any photos of the *DFS 228* or *DFS 346?*

MÖDEL - No, I have nothing. *Felix Kracht* had plans of both aircraft on microfilm at one time. You must ask him what happened to his microfilm. Nevertheless, I think that the Americans took it all from him after the war.

MÖDEL - You are aware of the American spy plane...*Flugzeug U-2*? I have never seen a picture of the *U-2* because at one time it was very secret.

MYHRA - Yes, it was. I'll send you a picture. It appears that the *DFS 228* was the father of the *U-2*. In this picture showing you the *DFS 228* pilot's couch, how old were you when the picture was taken?

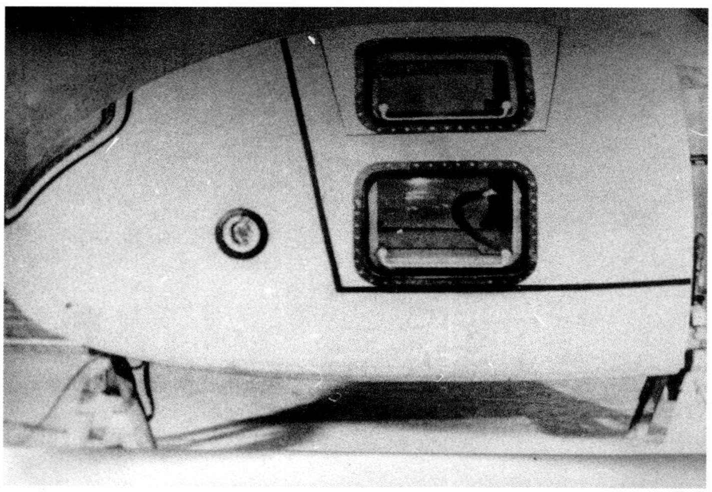

A direct port side view of the cockpit cabin for the *DFS 228 V2. Rolf Mödel* constructed this cockpit cabin out of wood according to *Felix Kracht's* design specifications. There are several changes from the *V1*. The fuselage side windows have been moved higher up on the fuselage wall. The pilot's entry and egress hatch extends down around the fuselage sides plus it carries a window on each side of the fuselage. In the nose we see that the round port hole on the *V1's* cockpit cabin has been changed to a more squarish plexiglass window. Finally, the swordfish emblem is more professional-looking due to being placed inside a circle.

The *V2's* cockpit cabin as viewed from its port rear side. The two metal tabs seen protruding on the rear wall of the cabin both top and bottom, are its attachment points to the center fuselage. Quick, disconnect connections can been seen about center on the fuselage rear wall.

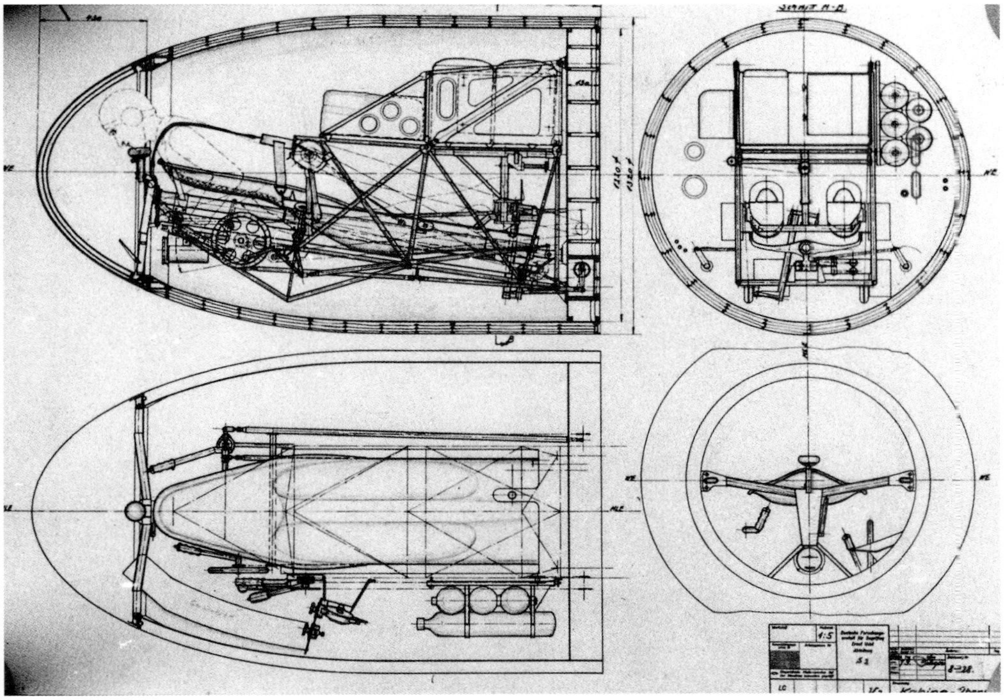

A pen and ink 3-view drawing from *DFS*-Ainring of their *DFS 228 V2* cockpit cabin. A major change between cockpit *V1* and *V2*, is that the pilot controls the aircraft while laying on a couch. The pilot's entry and egress was through the top hatch...like that on the *228 V1*, however the *V2's* upper hatch opening was larger. In addition, the pilot's couch was fixed inside the fuselage. The first cabin (*V2.1*) had two rectangular windows per side, however, this was changed in the second *(V2.2)* prototype cabin to only one window per side and that window was moved further aft than found on the *V2.1* cabin.

MÖDEL - That picture of me was taken sometime in 1943, and I was 32 or 33 years old. I was born in December 1910. We at *DFS* had learned a great deal from the *Berlin B-9* project and used many of its ideas in designing the pilot's prone/couch position on the *DFS 228/346*.

MYHRA - Really, could the *DFS 228* have been flown in 1945, with its *HWK 509D* rocket power?

MÖDEL - Yes, it could have been, but I don't know why exactly the *DFS 228* was not flown with its rocket operating. It was not due to the loss of cabin pressurization, even though cabin pressurization declined over time. The completed *DFS 228* had been taken to Ainring, and I remained at Hörsching. Our facilities at *DFS*-Hörsching were too small. I remember that a man by the name of *Rudolf Ziegler* was the pilot on the *DFS 228*. There were other problems with the cabin. I know that they had had difficulties at *DFS*-Ainring with humidity on the cockpit's large plexiglass window. They could not control its formation, and at *DFS*-Ainring they were working on a humidity absorber for use in the *DFS 228* and hoped to correct the problem this way. The engineer working on this problem was in opposition to the entire project. I don't know

The *DFS 228 V2* on the back of the *Do 217 K3* carrier aircraft. Notice what appears to be a single side window on the *228 V2*, but there is an outline of another window lower down on the cabin side wall.

why. We did not have problems with the plexiglass itself, but only with humidity forming on the plexiglass. But, yes, we still could have flown this machine with *HWK* rocket power, despite the cockpit problems.

MYHRA - When the Americans came to *DFS*-Ainring at war's end, what were their reactions to the *DFS 228 V1* they found there? Did these American officers personally talk to you?

MÖDEL - No, not me personally. So I don't know what they thought of our *DFS 228 V1* because the American officers didn't talk to me...only to *Felix Kracht*, because he was our section chief of all the *flugzeug* under development there at *DFS*-Ainring.

MYHRA - Were you invited to France after the war?

MÖDEL - Yes, I was invited to France in the Summer of 1946. Only three of us from *DFS* went to France: *Felix Kracht*, *Hermann Zitter*, and myself. *Eugen Sänger* and his wife went there, too. *Oberst Siegfried Knemeyer* went to France and worked there for a while, but got a better offer from the American government and he and his family left for Wright-Patterson Air Force Base.

MYHRA - Did the French also want to build a research aircraft like the *DFS 228* or *DFS 346?*

MÖDEL - Yes and no. Once we arrived they wanted us to make a design for a high altitude glider-sailplane for high altitude research similar to our *DFS 228*, but they later lost interest in this project and it was canceled. At first they didn't seem to want an *uberschall* aircraft like our *DFS 346* project. Instead, the French seemed to be more interested in a high altitude reconnaissance type of aircraft, and they wanted us to build a high altitude sailplane for that purpose.

Further mystery. This *DFS 228 V2*, atop its *Do 217 K3* carrier, has a *Balkenkreuz* (black and white *B3* style), but there also appears the single letter "*D*" aft the *Balkenkreuz*. Original, or added to the photograph later?

MYHRA - Do you have a photograph of this proposed high altitude photo reconnaissance sailplane?

MÖDEL - No, but they wanted the machine to carry instruments for measuring and carrying out high altitude photo reconnaissance. In fact, the French appeared to be very interested in all of our former *Luftwaffe* high altitude photo reconnaissance aircraft, techniques, and aircraft materials. They had asked us to help them better understand the use of such aircraft, materials, and instruments in gathering photographic intelligence.

MYHRA - What did *Dr.-Ing. Sänger* do for the French at this time after the war?

The *DFS 228 V2* on the back of the *Do 217 K3* carrier aircraft. Notice what appears to be a single side window on the *228 V2*, but there is an outline of another window lower The *DFS 228 V2* shown in gliding flight. Notice that a *Balkenkreuz* (black and white *B3* style) appears on the center section between the wing's trailing edge and the tail plane. It is not known if this *Balkenkreuz* is original or was added later to the photograph down on the cabin side wall.

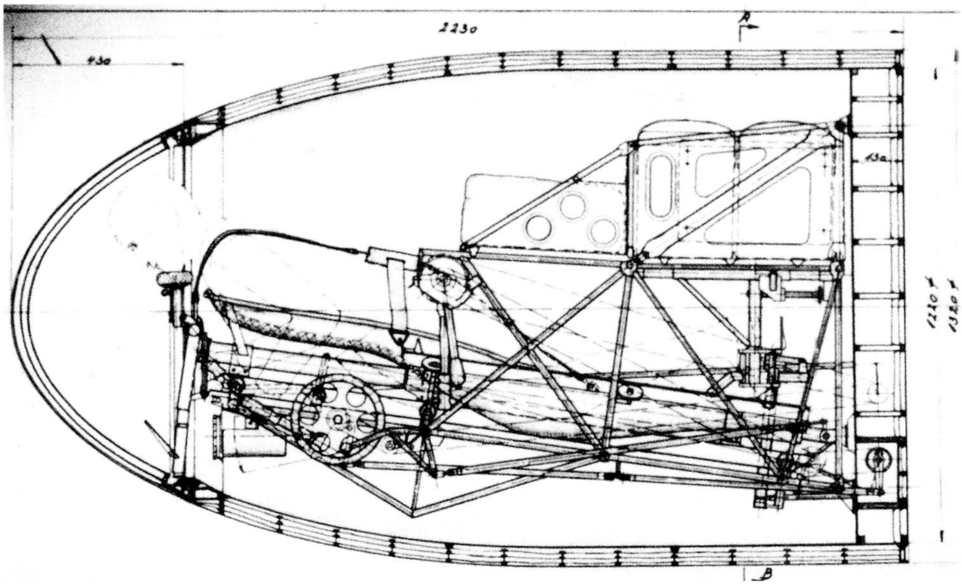

A pen and ink port side drawing of the *V2's* cockpit couch from *DFS*-Ainring. This couch was fixed inside the cabin. The pilot had to entry and egress via the cabin's top hatch.

MÖDEL - I don't know what he did for them, because he did not work in our department/group. He worked with a different French aeronautical group at another location.

MYHRA - Did the French ever express an interest in wishing to build an exact copy of the *DFS 346?*

MÖDEL - I think that they did later. I remember that the French became interested in an *uberschall* aircraft and some sort of rocket engine to power it. Then, too, they wanted to fly this aircraft without a pilot, however, like the English later did. The French wanted this research for guided missiles, I believe.

MYHRA - The French have kept some of these details secret, too, like the Soviets. The Soviets believed that the details of their speed of sound R&D was not necessary for other governments to know about. Who helped the French with their *uberschall* R&D?

MÖDEL - Well, the French were secretive about their post war aircraft research, and in some ways just as secret as the Soviets. Certainly the French had German help after the war on their *uberschall* activities. There was a German team of experts, and they were located in a research center in the south of France. I don't know much about this, but I recall that these German experts came from the different research institutes of the Technical High Schools and the universities throughout Germany. I don't recall them coming from the *Luftwaffe*, the *RLM*, or government-sponsored aeronautical centers such as *DVL*, *AVA*, *DFS,* and so on. I believe that when *Dr.-Ing. Sänger* worked for the French he worked on *uberschall* research, and he was also located in the south of France.

MYHRA - What other individuals are available to talk to about the *DFS 228/346?*

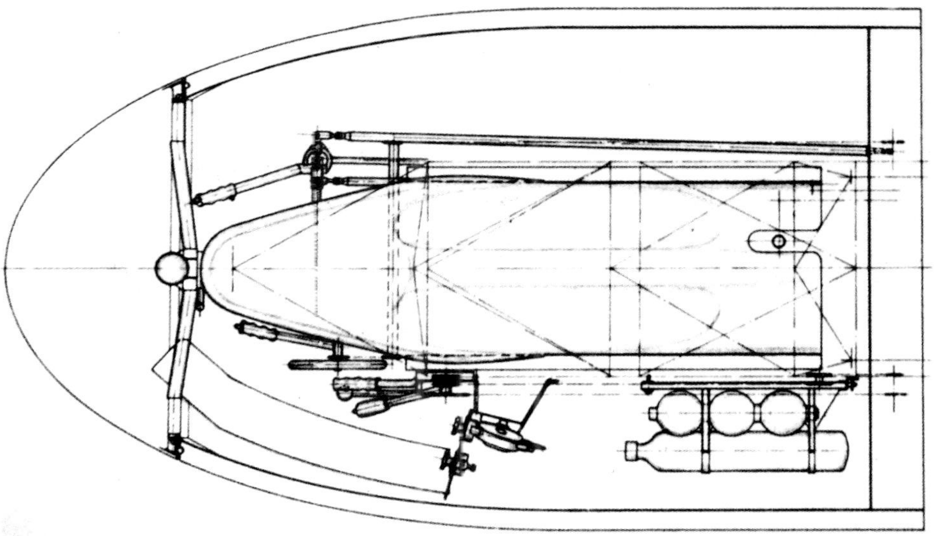

A pen and ink top view drawing of the *V2's* cockpit couch from *DFS*-Ainring. In the lower right hand corner are oxygen bottles...round ball-like ones for the pilot and more conventional cylindrical ones to pressurize the cabin.

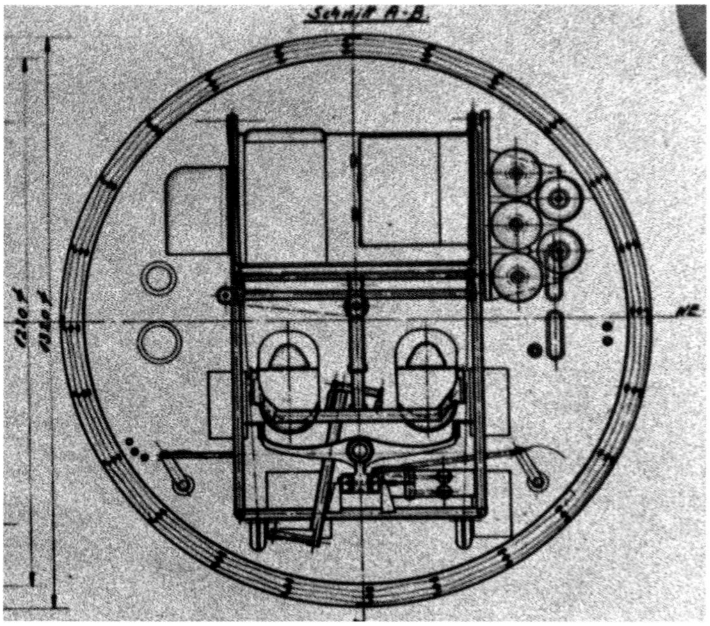

A poor quality pen and ink cockpit cabin cross section drawing of the *228 V2* from *DFS*- Ainring. What we see is the rear of the cabin looking forward toward the nose. Oxygen bottles appear on the starboard side of the cockpit.

MÖDEL - You could talk to *Professor Niese*, *Professor Eberhard Meyer*, *Hans Zitter*, and *Klöckner*.

MYHRA - I've talked to each of them already. At *DFS* at that time *Niese* was working mainly on the intermittent pulse jet powered *Messerschmitt Me 328*, a research interceptor aircraft. He told me that he was not that close to either the *DFS 228* or *DFS 346*.

MÖDEL - Yes, that is correct. I flew the *Me 328*, but only as a sailplane and never with power.

MYHRA - Do you know if a scale model of *the DFS 228* or *346* had been tested in a wind tunnel at any time?

MÖDEL - Yes, both had been wind tunnel tested, but I don't know where or when.

MYHRA - I've heard that *Professor Quick* had tested a scale model of the *DFS 346* in a wind tunnel at *DVL*-Berlin/Adlershof.

MÖDEL - Yes, and, by the way, *Professor Quick* was also in France. I remember, too, that *Quick* worked on the *HWK 509D* bi-fuel liquid rocket engines used in both the *228* and *346*.

End of *Rolf Mödel* Interview

Eberhard MEYER
29 May 1987
Heiligenberg, West Germany

MYHRA - I'm researching the history of the *DFS 228* bi-fuel liquid rocket-powered ultra high altitude reconnaissance aircraft.

MEYER - I worked on this project at *DFS*-Ainring all through 1944. *Professor Georgii* was the director of *DFS*, and *Felix Kracht* was my boss. Before Ainring, we were all located at Darmstadt. *Hanna Reitsch* also worked for the *DFS* at Darmstadt, too. Later we all moved to Ainring, near Lake Chiemsee. At that time, too, there was a Technical High School near Chiemsee (between the cities of Munich and Salzburg) with a department for aeronautics. The students in this department of aeronautics were able to benefit when *DFS* moved to Ainring.

Several scale models of the *DFS 228* were built, but in the form of a two seat sailplane, and their aerodynamics were tested in strong winds (*Phron* winds) which blow in the Alps mountains between Salzburg and Austria in the spring. *DFS* had a small house at this site, and the idea was to carry/pull our sailplanes up to this house on the mountain and then fly them down. These small sailplanes were to be equipped with oxygen for the pilot, because the aircraft was flown to a height of 111 kilometers [71,461 feet]. These sailplanes were also to be equipped with a pressurized cockpit/cabin. Even before the war started we had built such a sailplane to test pressurized cockpits for use in sailplanes.

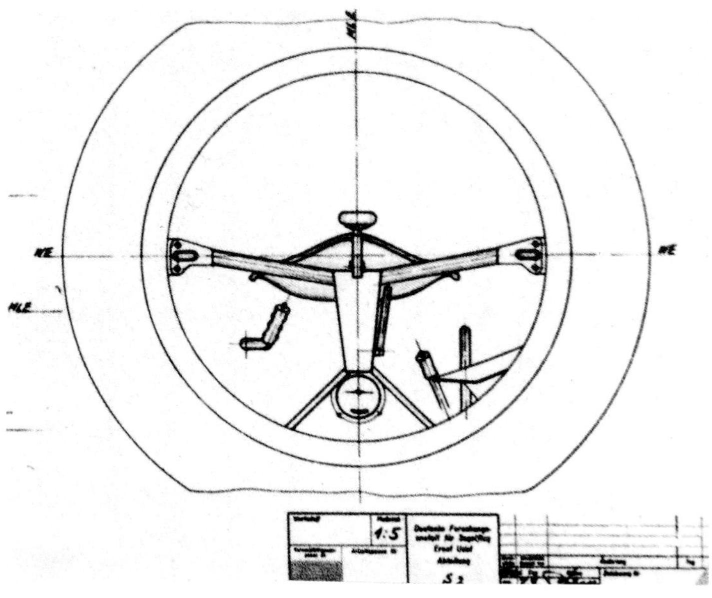

A pen and ink cockpit cabin cross sectional drawing of the *228 V2* from *DFS*-Ainring featuring the front of the cockpit cabin looking aft. In the center of the drawing is the pilot's chin rest which is built into the bicycle like handle bar main flight control.

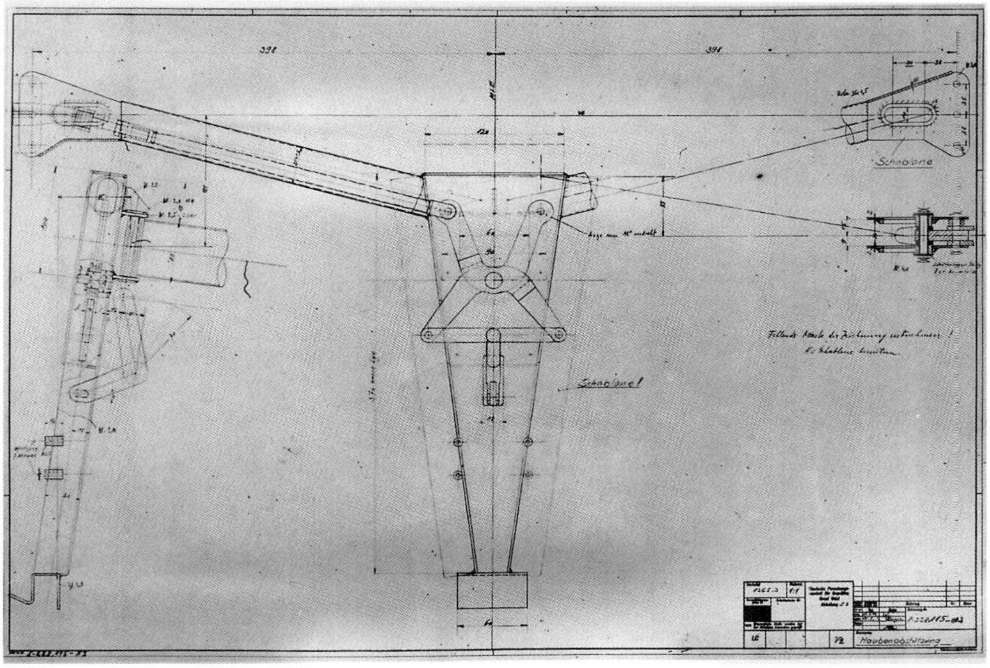

A pen and ink drawing of the *228 V2's* handle bar control apparatus from *DFS*-Ainring. The *228 V2's* cockpit cabin with the prone piloting position was designed by *Kracht* and *Meyer* and built by *Mödel*.

MYHRA - *Felix Kracht* told me about two of his sailplane friends who flew *Horten* all-wing sailplanes. *Heinz Scheidhauer*, who had suffered severe frost bite, and *Walter Blech* had died during the *Rhön/Wasserkuppe* summer sailplane competitions of May 1938, when their sailplanes were taken up on a thermal. *Kracht* said that its goal was to design a pressurized cockpit for sailplane use and thus prevent any more tragedies like those suffered by *Scheidhauer* and *Blech* during the August 1938 Summer Sailplane Competitions at the *Rhön/Wasserkuppe*.

MEYER - Yes, this is so. Our first pressurized cockpit cabin prototype *DFS 54* was initially built out of wood, and we wanted to test it up to 12 kilometers [74,568 feet]. The idea was to take this sailplane up in the spring winds around Salzburg and then reach 12 kilometers. But when it was ready it was now 1939, and with the war beginning in September 1939, *DFS*, in a civilian capacity, was not allowed to make these high altitude sailing flights anymore.

MYHRA - But *DFS* continued to work on the pressurized cockpit cabin in the *DFS 228*, off and on, throughout the war?

MEYER - Yes. The *DFS 228* pressurized cockpit cabin was the forerunner of the *DFS 346*. The *346* would require such a cabin were it to break the speed of sound. Like in the *228*, the pilot in the *346* also would pilot the aircraft while lying in a prone position on a couch in a pressurized cabin. So you see, the *228*, which we constructed at *DFS*-Ainring, was the forerunner of the *346*, that is, it was but one step before our *346*. But since it was then wartime, *DFS* was not allowed to work on civilian sailplane projects any longer.

MYHRA - This meant that *DFS* had to concentrate only on military related research?

MEYER - Yes. But we still wanted to fly the *DFS 228* and then go on and build the *346* so we could fly at or beyond the speed of sound. At the beginning of the war in 1939, the aircraft flown by the *Luftwaffe* flew at a relatively slow speed of between 400 and 500 mph. However, in a power dive, the *Bf 109* could reach 900 km/h [560 mph], and the *Me 163* could reach 1,000 km/h [621 mph] forward speed. When these aircraft approached *Mach 1.0*, especially the *Me 163*, they were unable to come out of the dive and tended to crash into the ground instead. The *Me 163* was not con-

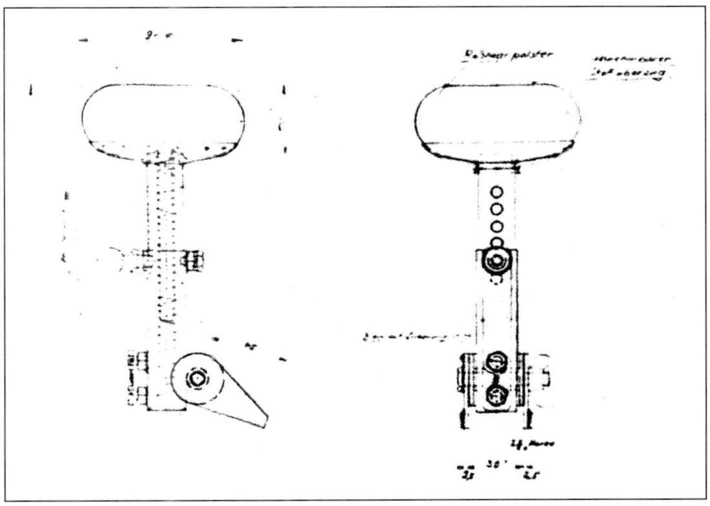

A poor quality pen and ink drawing of the pilot's chin rest, from *DFS*-Ainring, as it appeared in the *228 V2*. The upper round object is the cushion where the pilot placed his chin for support.

structed to fly at a rate of 1,000 km/h. At these speeds the rudder would not work correctly, nor would the other controlling surfaces. As a result, the aircraft could not be pulled out of its dive and would dive into the ground.

MYHRA - We had the same experience in the USA, too. One way in which the lack of effective control was corrected was through what is called an all moving tail, or a flying tail.

MEYER - The well known *DFS* pilot *Hans Zeigler* died when he could not recover his aircraft while in a power dive. Early on *Udet* had a great deal of interest in finding ways of bringing high speed fighter aircraft out of a power dive. As a result, the *DFS* wanted to test what type of material, metal, and so on were necessary to safely fly at the speed of sound without resulting in a crash. *Georgii* of *DFS* suggested to the *RLM* that they be allowed to construct a research aircraft to be able to test different features, metals, and planforms that would result in a safer high speed aircraft. *DFS* maintained that it was very important to them to develop a safe high speed *Mach 1.0* airplane.

***DFS*-Ainring engineer *Rolf Mödel* is trying out the *228 V2's* pilot's couch which he has constructed from plans drafted by *Kracht* and *Meyer*. The couch could be moved forward or aft, depending on the height (length) of the pilot, so that he could comfortably reach the foot operated pedals. His left hand is on the trim wheel. The stick near the trim wheel, with the black nob was the control stick for elevation trim.**

MYHRA - So it was *DFS'* idea to construct a speed of sound type of research test aircraft?

MEYER - Yes. My immediate boss at *DFS* was *Felix Kracht*, and he was the leading person in *DFS* pushing for the construction of first the *228* and then the *346* as research efforts to perfect safe high-speed planforms. *Georgii*, too, became a supporter of finding a safe *Mach 1.0* airframe. I, as *DFS'* leading aerodynamicist, was responsible for soliciting ideas for such an aircraft from leading people in the aviation industry, the research labs, and universities.

On our *DFS 228*, and later with the *DFS 346*, we designed the fuselage into three sections: the detachable cockpit or cabin; the center section, which housed the fuel tanks for the bi-fuel *HWK 509* bi-fuel liquid rocket engine; and the tail section, which contained the thrust nozzles—two in the *346*, while only one in the *228*.

The prototype pilot's couch as it appeared in the *DFS 228 V2,* as seen from its nose front port side. Notice the pilot's chin support in the upper left hand corner of the photo. Across the bottom of the photograph can been seen a bank of switches. They were electric starting and running switches. At the upper right hand corner of the photo are instruments. The pilot would view these instruments via a mirror. The main flight instruments were beneath the pilot's chin and in clear view.

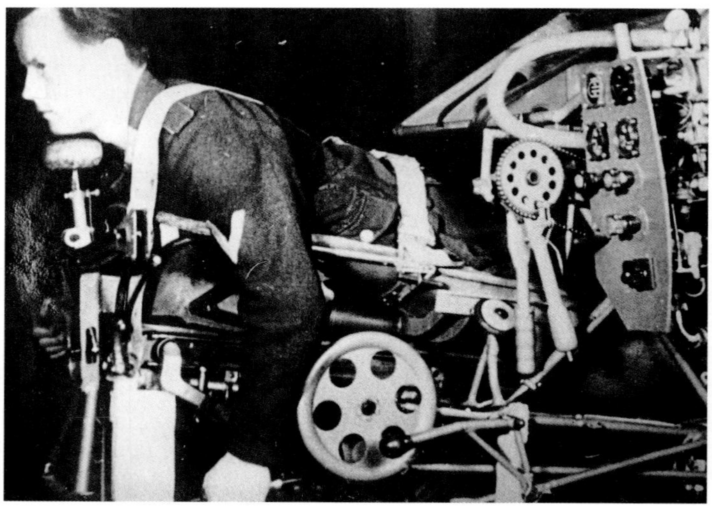

***Rolf Mödel* is shown with both hands on the *V2's* flight control bicycle-like handle bar. Notice, too, that *Mödel's* upper legs, knees, and feet are being supported by a fabric pouch visible in the photograph below the instrument panel.**

The essential cockpit cabin instruments for the *228/346* are featured in this photo. They have been installed beneath the pilot's chin rest and out forward so that the pilot only has to move his eye downward to read their information.

A close-up view of the port side of the pilot's couch in the *228 V2*.

In the *228/346* a large parachute was placed in a box near the rear bulkhead of the cockpit cabin. The cabin itself was attached to the fuselage center section by four explosive bolts, allowing the cabin to break away from the aircraft in case of an emergency. The *346* was designed to reach 30 kilometers [98,430 feet], but our cockpit design did not work out as planned. The continuous problem we had was that we could not make it tight enough and thus keep it pressurized at altitude. Anyway, *Hans Ziegler* died as a result of the failed testing.

MYHRA - How was the aircraft able to reach a height of 30 kilometers?

MEYER - By a mother aircraft, usually a twin-engined *Do 217K*. If an emergency occurred at 2 kilometers [6,562 feet], for example, the pilot pushed another button which took him out of the cockpit cabin so that the pilot could land normally. Of course, the pilot could not do all this at 30 kilometers, because he would freeze to death at that altitude.

In the center or mid-section on the 228/346 were located the bi-fuel rocket engine fuel tanks. Also located in this mid section were numerous stress measuring gauges. The idea was to measure the stress occurring on the aircraft at 32 different points. In the *DFS 346* were four recording devices, or machines. The data was then transmitted via radio signal to a ground control radio receiving sta-

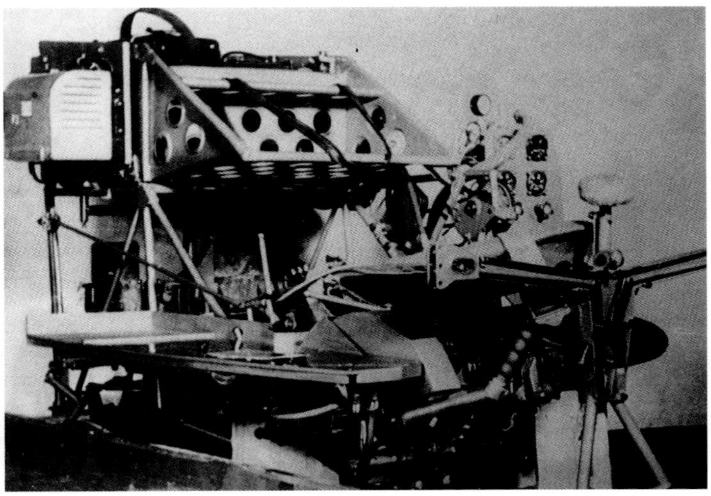

The entire prone piloting couch for the *DFS 228 V2* as viewed from its nose/front starboard side. The metal rack above the couch is where the pilot's parachute was stored. Notice at the far right of the photograph, the pilot's chin rest and beneath it, the bicycle-like handle bar for main flight control.

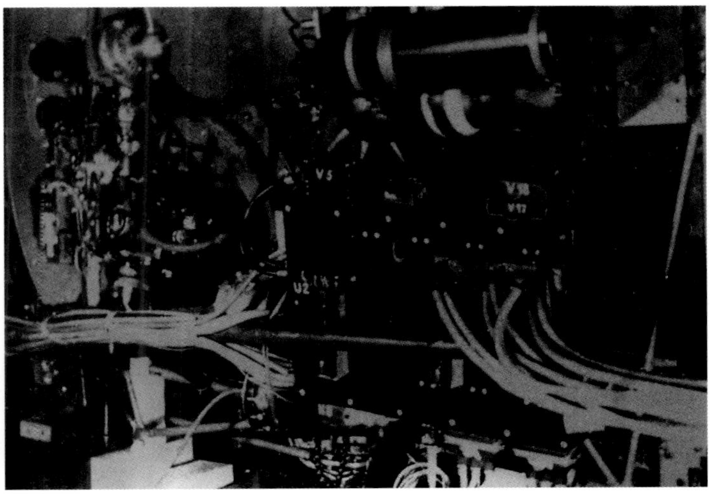

A view of the pilot's couch in the *DFS 228 V2* as viewed from its rear port side looking forward to the cockpit plexiglass window. Notice the stack of oxygen bottles in the upper center of the photograph. They were used for the pilot's oxygen and air to pressurize the cockpit. A flight instrument panel for secondary instruments, viewed through a mirror by the pilot, is located in the upper left hand corner of the photograph.

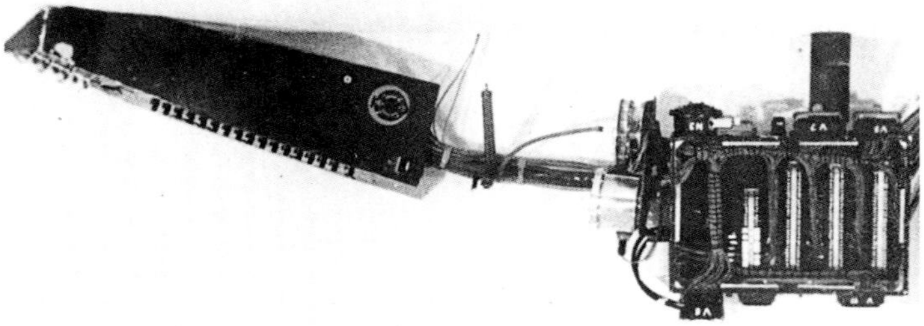

To the left of the photograph is a bank of switches used by the pilot in the *DFS 228 V2*. To the right is the electrical control box.

The control box, its wiring coming from the bank switches seen above.

tion. The wings were attached to the mid-section. In the aft, or third, section of the fuselage was housed the special *HWK 509* bi-fuel rocket engine.

In addition to the *DFS 228* being used as a research vehicle for testing its pressurized cockpit cabin at high altitude, a short series production would have been used as a high altitude reconnaissance aircraft. It was designed to reach 23 kilometers [75,000 feet] in height. So, a series production of *228s* were planned, and each one would have been constructed entirely out of wood.

MYHRA - The idea for the 346 came from *DFS*. Once it was approved, how much input did *DFS* receive from the *RLM*, and since they financed the *346*, too, what requirements did they have, if any? When was it to be completed?

MEYER - The idea for the *346* originated between me and *Kracht*. We developed the planform for the *346* out of our experience with our *DFS 228* powered sailplane design. No one else was involved at this early idea stage. No, we did not even receive any input from the *RLM* on the *DFS 346's* design at this time, either. In effect, I recall, we told the *RLM* that we wanted to construct a research vehicle like this, and the *RLM* people we knew, such as *Knemeyer*, *Antz*, *Riedenbach*, *Schelp*, and so on, all said in effect "yes, go do it." But, usually when the *RLM* gave us money to carry on research, they had a right to demand changes. But in the case of the *346* this did not happen. We at *DFS* had no deadline by which the *346's* design had to be completed nor constructed. *Professor Georgii* liked and approved the idea for the *346*. Together, *Georgii*, *Kracht*, and I went to the *RLM*-Berlin and told our supporters about our evolving plans for the *DFS 346*. After the idea was presented and interest had been generated, *Kracht* and I would go frequently to Berlin, together. Sometimes I'd go there by myself. Naturally, we had to report to the *RLM* our progress on the engineering drawings. Progress on the *346*, especially in those desperate days, went along quite fast. Nevertheless, when one is designing a research aircraft like the *346*, one cannot be in a hurry, anyway. The *RLM* understood this fact, and they did not push us to complete the *346* by any set date.

MYHRA - The man in charge of the *RLM*'s Technical Office was *Oberst Siegfried Knemeyer*.

Minor flight control instruments at the rear of the pilot's couch, in reverse lettering, which he viewed in a mirror.

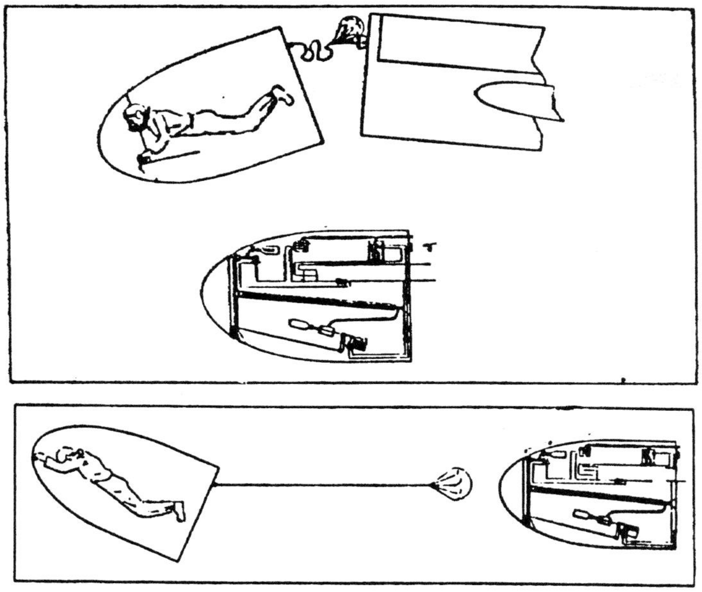

A pen and ink drawing from *DFS*-Ainring featuring the pilot escape mechanism built into the prototype cockpit for use in the *DFS 228 V3*. This prototype under construction by *Mödel* at war's end from plans provided by *Kracht* and *Meyer*. High technology in 1944. The entire cockpit cabin was separated from the *228 V3's* center fuselage by explosive bolts invented by *Dip.-Ing. Karl Bütter* of *Heinkel AG*.

MEYER - Yes, but I do not remember anymore, at this time of my life, which person(s) was most interested in having the *346* built. This has now been almost fifty years. However, I do remember *Knemeyer* from the *RLM*. *Udet* had been a strong supporter of our work at *DFS*, and we would often meet and talk with him. After *Udet's* death on 17 November 1941, we then would meet with *Milch*. Later on we moved our *DFS* research department/offices from Darmstadt to the *Flugplatz*-Ainring near Salzburg.

MYHRA - Why were your *DFS* departments moved to Ainring from Darmstadt?

MEYER - By this time of the war in early 1944, the war was not going good for Germany. Darmstadt was not too far from the French border, and should the Allied troops overrun Darmstadt we did not want the Americans or French to capture all our documents and research material. In fact, all the higher ups moved to more safe locations deep in Germany.

MYHRA - Where did some of these innovating design ideas for the *DFS 346* come from? For example, its T-tail. Did this idea come from *Hans Multhopp*, for example?

MEYER - No, the T-tail idea was my own idea. I suggested it and it was accepted. The reason for the T-tail had to do with potential turbulence coming in the wake of trailing edge of the wing at high speed, and I believed that this turbulence would adversely affect the *346's* controlling surfaces on its tail plane.

MYHRA - It appears that *DFS* assembled a small group of aeronautical people on the *DFS 346*. These men included *Hans Multhopp*, *Siegfried Günter*, *Rüdiger Kosin*, *Woldamar Voigt*, the *Göthert* brothers, *Professors Quick*, and *Busemann*. How much influence did these men have on the final planform of the *DFS 346?*

MEYER - Not very much. We at *DFS* felt that it would be wise to have these leaders of the aviation industry serve as advisors on this high speed project. We wanted to pass along our ideas, the type of aircraft which we at *DFS* had in mind, to them for their comments. In other words, we felt it was good insurance for us to have these people on our side should someone start opposing the *DFS 346* idea. Besides, as leaders of the industry, we could gain valuable knowledge for the *346* project. Our work at *DFS* and our ideas on the *346's* proposed planform turned out to be very frustrating. For example, I had the idea to use a T-tail on the *346*. When this idea was presented to our small committee, well, *Multhopp*, for example,

Left to right: *Ernst Udet*, *Karl Bütter*, and *Paul Bader* of *Erprobungsstelle*-Rechlin. About 1940.

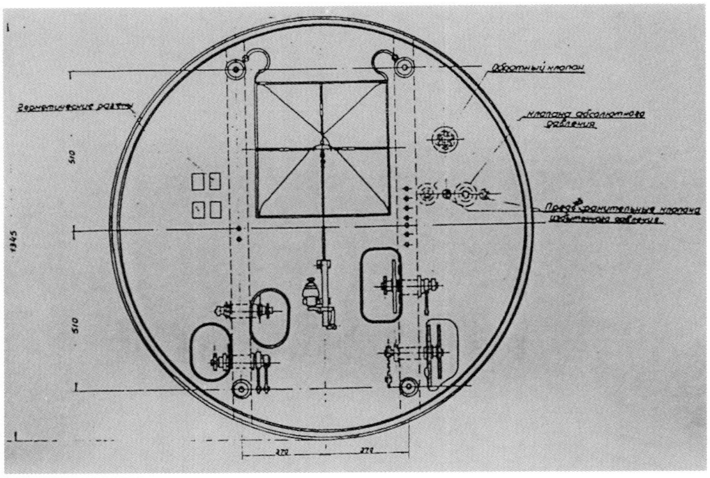

A pen and ink drawing of the 3rd *228/346* cockpit cabin's rear bulkhead and sealing frame as viewed from its stern. The square box inside is the storage location for the cabin's stabilizing parachute.
right top - hermetic connections
left top - down
- reverse valve
- absolute pressure valves
- excess pressure safety valve

said, "Hey, I was going to suggest this tail arrangement, too, this is the same idea I had." *Rüdiger Kosin* said that it was his idea to give the wings a 45° sweep-back, and so on. Well, most of the ideas which *Kracht* and I had for the *346*, well, one or more people on our advisory committee claimed that they had had a similar idea. Naturally, then, they would take all the credit for their idea, after we had presented it! That was so frustrating to us. But this advisory committee was a tradition which we had to honor...for several reasons. But we gave this committee all our ideas for a supersonic research vehicle. They in turn told us "it was correct," and that they were happy that we had taken all "their" new ideas and placed them in the *346's* planform.

MYHRA - Tell me about *Professor Quick* of *DVL* and the *Göthert* brothers? There were two brothers...one from *Gothaer* and the other from *DVL*. The two brothers, *Bernard* and *Rudolf*, were both strongly into theoretical aerodynamics. At one time they worked together at *DVL*. Then, in late 1943, *Dr.-Ing. Rudolf Göthert* joined the *Flugzeugbau* department of the *Gothaer Waggonfabrick* in the City of Gotha and worked on the *Horten Ho 229*, plus an advanced *Gothaer* version of an all-wing fighter called the *Gotha Go P60 series A, B, and C. Göthert* still maintained close ties to *DVL* and was able to use their wind tunnels with the help of his brother.

MEYER - *Professor Quick* was from the *DVL*-Berlin/Adlershof, and so was *Rudolf Göthert*. We wanted the *DVL* in on the *346* project because we then could use their wind tunnel. We had no wind tunnel at *DFS*-Ainring, so with *Professor Quick* and *Dr.-Ing. Göthert* of *DVL* we were able to gain access to their wind tunnels and exchange experience. *Quick* and *Göthert* placed a scale model of the *346* planform in their wind tunnel, and in turn gave us valuable data which either confirmed our ideas or made us redesign certain parts. Their input was most valuable to us on this project.

In addition to the airport at Ainring, there was a second airport used by the *DFS* east of Munich by the name of Prien (next to Lake Chiemsee on Route 305 and across the *Autobahn* from Bernau),

The real thing! The 3rd cockpit cabin for the *DFS 228* and the later *DFS 346*. It was never connected up to a *228* fuselage, however, it was post war, in the USSR, attached to the *DFS 346*, flown, and used as an escape mechanism for high flying *Flugkapitän Wolfgang Ziese*.

Testing of the 3rd cockpit cabin prototype intended for the *DFS 228* and *DFS 346*. The unmanned cabin was taken up to altitude via a carrier aircraft and released. In this poor quality photograph, we see the cabin moments after being released.

The 3rd *228/346* cockpit cabin's parachute is seen deploying in this poor quality photograph.

The 3rd *228/346* cockpit cabin prototype has reached a safe altitude as detected by sensors and the pilot's couch is being ejected nose first out of the cabin. The pilot's couch, as can be seen in this poor quality photo, has had its own parachute deployed.

and students from the Technical High School of *Munchen* would come out to the Prien airport. These students' constructors/pilots were good, and several from this Technical High School had flown over the Alps in their own aircraft. Some even went so far as Italy.

MYHRA - How did it come to be that *Multhopp* was a member of the *346* advisory committee?

MEYER - *Hans Multhopp* was an aerodynamicist from the *flugzeug* firm of *Focke-Wulf*-Bremen. He had been a former student of one of the Technical High Schools. *Professor Georgii*, on the other hand, was too old to fly anymore. But *Hans Multhopp* had proven himself to be a good man, and had gone to work for the *DVL* while still a student. These students generally received money from the *DVL* to help them in their studies and their school aircraft building projects. Then, upon graduation, the students would come and work for a time at *DVL*. The *DVL* also provided tow planes for towing the students' gliders into the air. Other students who were not so good were hired as carpenters, and so on.

MYHRA - In what year did the active design thinking start on the *DFS 346?*

MEYER - On the *346* in early 1942. But the *DFS 228* was thought about as early as 1939. Its single *HWK 509* could be used to obtain a great altitude of 23 kilometers [75,000 feet]. I flew one of the *DFS 228s*, but without power. Since the *228* was built out of wood, it would have been very hard to be detected by radar. I believe that the American's idea for the *U-2* was taken from our *DFS 228!*

MYHRA - *Reimar Horten* had the same idea with his *Ho 9* all-wing fighter too. Why did you start thinking/construction on the *DFS 346* so late in the war?

MEYER - Because the overall speed of aircraft at this time, thanks to the turbojet and rocket engines, had increased significantly from the speed at the beginning of the war. We needed research data on speeds beyond sound. For example, the engineering people at *Messerschmitt AG*-Augsburg who worked on the *Me 163* were very interested in our *346* idea. The *Me 163* would reach 1,000 km/h [621 mph] regularly in a power dive. This was almost *Mach 1.0*. At this time, too, *Messerschmitt AG* was working on a twin-

The pilot's couch/bed parachute is seen in the far left of the photo fully let out and in a few moments its stabilizing parachute will be completely deployed. The empty cockpit cabin is seen to the far right in the photo as it now begins its free-fall back to earth.

chamber *HWK 509* interceptor called the *Me 263* (It had been started by *Junkers*-Dessau. At this time it was known as the *Ju 248*, but then it was given over to *Messerschmitt AG* to complete and redesignated as the *Me 263*). It was possible that the *Me 263* could easily exceed *Mach 1.0* in a power dive, so the people working on the *Me 263*, such as *Voigt*, were very interested in our *346* project.

We had an additional problem with high speeds. For example, due to the increased speeds of our interceptors, like the *Me 163*, when they were making an attack on a *B-17* bomber formation from altitude they would reach speeds near the supersonic level. Then, after the attack the *Me 163* would have a difficult time recovering from this high speed in order to make a second pass. Some *Me 163s* could not recover, especially if they were forced into a dive to evade Allied fighters. They ended up diving right into the ground! This was always a problem when the *Me 163* came too close to *Mach 1.0* speeds. It was becoming a big problem in trying to shoot down the *B-17* bombers with the *Me 163*. It would have been even more so with the *Me 263*. Anyway, apparently the Allied fighters knew that if the *Me 163*, even the *Bf 109*, were forced into a high speed dive then there was the possibility that it could not recover and it would crash. See, an easy way to kill our fighters! So, the high speed of the *Me 163* was a grave disadvantage in one way, just as its high speed also had great advantages.

I went frequently to *Messerschmitt AG* in Augsburg to talk with *Woldamar Voigt*, because he especially, and others, as well, at *Messerschmitt AG* were very interested in learning how to bring high speed aircraft who had nearly reached *Mach 1.0* in a power dive out of this dive. High speed was becoming a grave problem for us, and it appeared that the problem would increase as our turbojet powered aircraft types increased. We, therefore, had to find out how *Mach 1.0* flights, especially in a dive, could be made safe through better airframe design and control. This is the reason why we started so late in the war. The problem did not exist at the beginning of the war, but only appeared by 1943 and 1944, as the fighters got faster through higher powered engines. At the beginning of the war in late 1939, our fighter speeds were between 500 km/h [311 mph] and 600 km/h [372 mph], and we had no problems even in a power dive with this speed. Then, with the use of the *Me 163*, 1,000 km/h [621 mph] was routine in a power dive. So, we needed now to find out what caused these control problems at speeds around *Mach 1.0*, and then what we could do to correct it. This was the only reason why we at *DFS* started so late in the war to do research into *Mach 1.0* speeds. Pilots as well as designers were frustrated because they said if our aircraft can reach speeds of 1,000 km/h [621 mph] and higher with all these new engines, what can we do to prevent flight control problems at these speeds. There was considerable interest in safe handling at 1,000 km/h and higher speeds. The industry wanted us to find out what caused the problem and how to correct it. Plus, this issue of flight control at supersonic speed was of considerable interest to all the aviation departments

***Karl Bütter*, Munich, West Germany. September 1988. *Bütter* had the pleasure of being taken captive by the Soviets and sent to the USSR (1946-1955) to help prepare the *DFS 346* for flight testing. Photo by the author.**

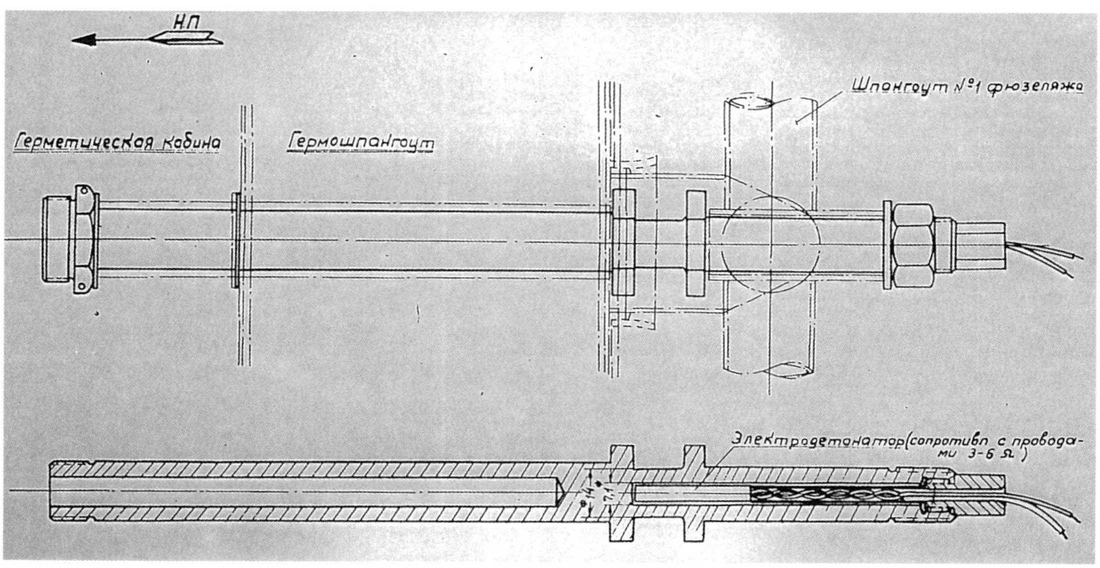

A pen and ink drawing of *Karl Bütter's* explosive bolt used to separate the cockpit cabin from the center fuselage on the 3rd *228/346* cockpit cabin. It was originally drawn by *DFS* but traced and copied by the Soviets post war. Upper pen and ink drawing left to right:
- **hermetic cabin**
- **sealing frame**
- **fuselage frame #1**

Lower pen and ink drawing left to right:
- **electric detonator (resistance with wiring 3-6 ohm)**

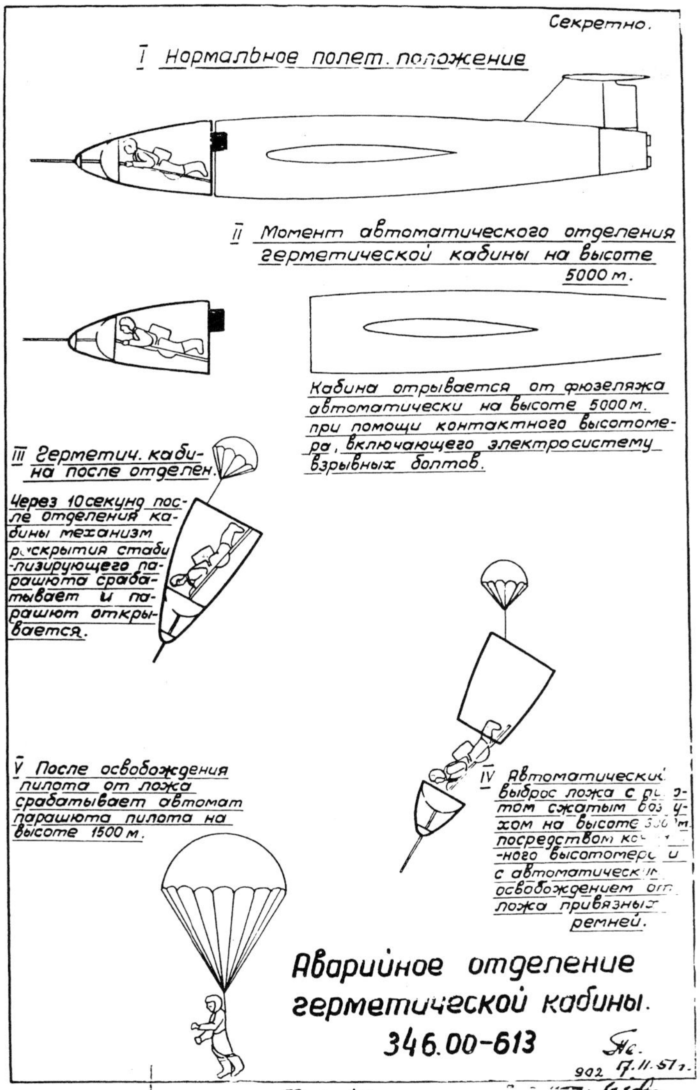

Фиг. 43

DFS' prototype 3rd prototype *228/346* cockpit cabin and the procedure used to save the high-flying pilot at an ultra high altitude. During an emergency, the entire cockpit cabin is separated from the center fuselage section by four explosive bolts. When the cabin is free of the fuselage, its stored parachute is automatically deployed. When the cabin with pilot still in, reaches a prescribed safe altitude, the pilot still belted to his couch are ejected. The pilot and couch are stabilized by its own parachute. At a prescribed safe altitude, the pilot, free of his couch, parachutes to earth with his own parachute.

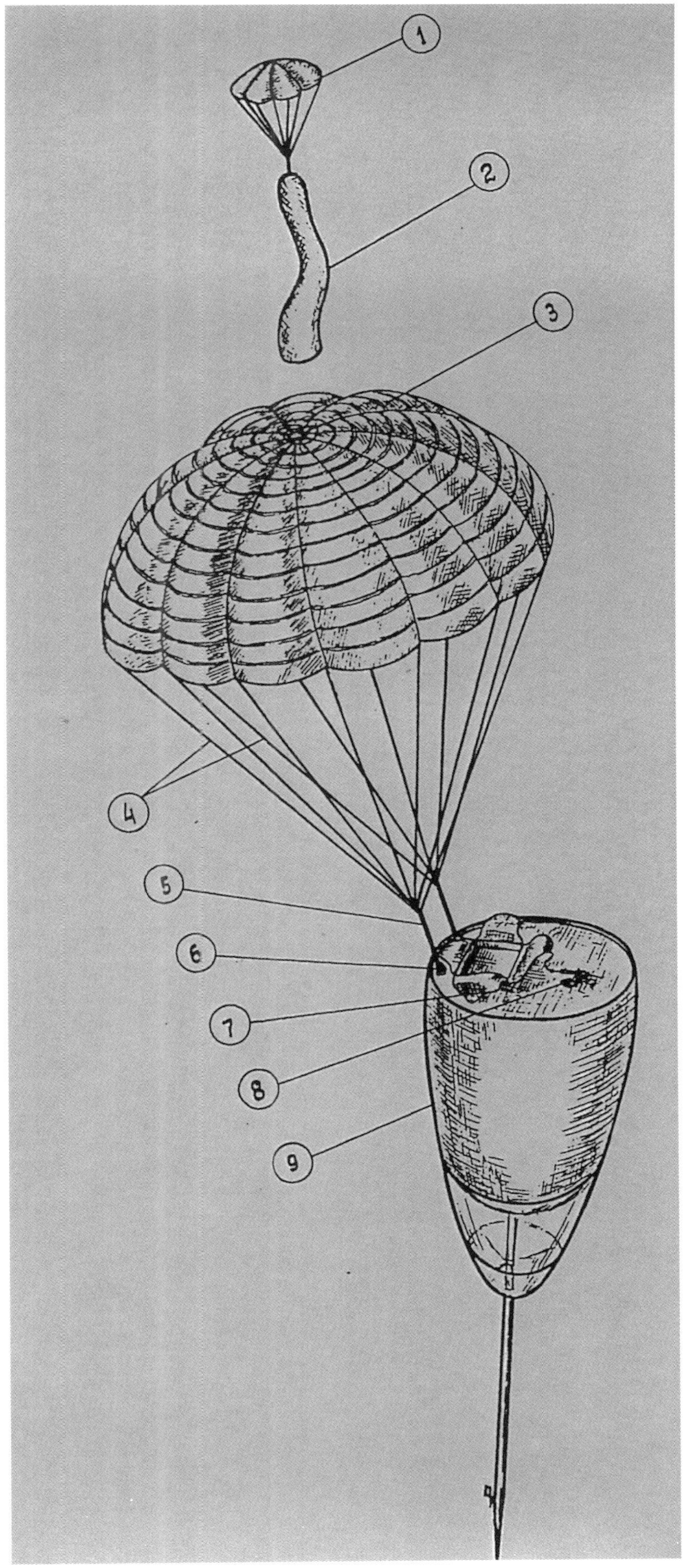

Right: A pen and ink drawing of he *DFS'* 3rd prototype cockpit cabin shortly after being blown free by explosive bolts and being stabilized by its own parachute.

01 - pull-out parachute
02 - parachute storage stocking
03 - stabilizing parachute
04 - cords
05 - cable
06 - cords' fastening points
07 - parachute knapsack
08 - deployment device
09 - hermetic cabin

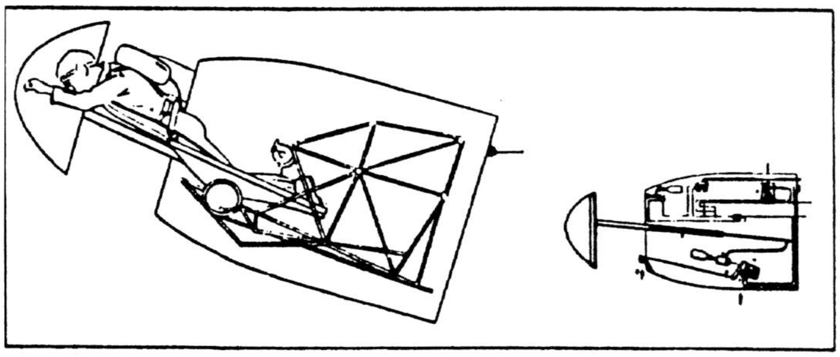

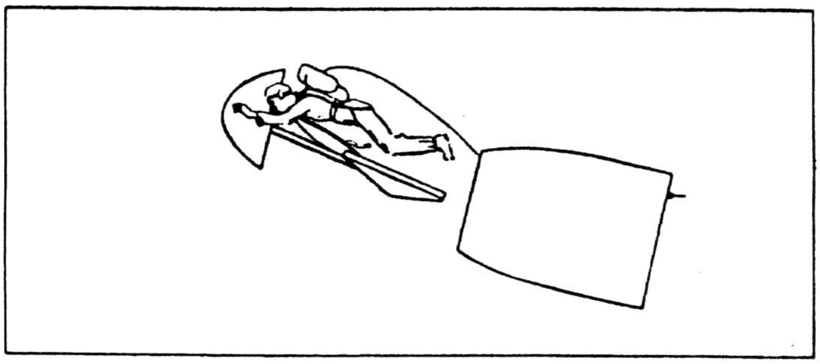

A pen and ink drawing by *DFS* featuring the sequence of events when the pilot and couch are ejected out of the cockpit cabin. In the first stage, the pilot still strapped to the couch are pushed out of the cabin but still protected by the plexiglass nose covering.

in the various Technical High Schools throughout Germany. This was because in these aviation departments were the students who would be the next generation of aircraft designers in Germany. It was these students who would be into high speed aircraft powered by the new turbojet engines. I'm an engineer, but I was also a test pilot, and on occasion I flew most of the fighter aircraft built in Germany during the war. These aviation students would frequently come to *DFS* and *DVL* to see for themselves the new planforms under development. Well, we all knew each other, those of us in aviation design and development throughout Germany, so we could exchange ideas. But although I knew people like *Hans Multhopp*, I did not know him very well.

MYHRA - How were these committee people selected...people like *Multhopp*, *Kosin*, *Voigt*, and the others?

MEYER - They were the leading aerodynamic research people in industry, that is, people who were pushing toward *Mach 1.0* speeds.

MYHRA - How did it come to be that *Siebel*-Halle was selected to construct the *DFS 346?*

MEYER - Well, we in the *DFS* did not have the ability to build aircraft out of metal, so we needed a *flugzeugbau* in order to construct our *346. Messerschmitt AG*, *Focke-Wulf*, and *Henschel* had no extra capacity at the time of the war. On the other hand, *Siebel* had capacity, and therefore was the only *flugzeugbau* able to build the *346.* Other firms were also suggested/asked, such as *Heinkel AG* but he, too, had no capacity. So *Siebel*-Halle was the choice.

There was an additional problem due to the fact that we only wanted two *346* prototypes built and not 2,000. One of a kind research aircraft, like the *346*, are more difficult, more time consuming to construct than series production aircraft. Thus, *Siebel* was thought to be the most suitable: free capacity, experience in metal working, and a willingness to build two prototypes...the first try. Plus *Fritz Seibel's wife was the sister of Hermann Goring's* wife.

The French also obtained the drawings for the *DFS 346* immediately after the war. In fact, the French took many *DFS*-Ainring workers to Paris, such as *Kracht*, *Nestler*, *Georgii*, *Sänger*, *Zborowski*, and *Mödel.* They tried to continue with the construction of the *346* in France, but shortly after they started they discovered that its construction and development would be too expensive for them.

MYHRA - What type of flight testing program was envisioned for the *346?* Where would this flight testing have been carried out? At *Erprobungsstelle*-Rechlin or *Siebel*-Halle?

MEYER - First of all, our *346* would have been taken aloft to 10 kilometers [32,810 feet] by a bomber aircraft and then released. Upon release it would have flown under its own power to 30 kilometers [98,430 feet]. Unlike in America where you have these great dry salt lake beds on which super fast aircraft can land, like Edwards Test Center in California, we have none of these in Germany. So, we would have had to find a very long, suitable *flugplatz.*

MYHRA - Where in Germany at this time (1945) would you have found a useable and very long runway? Since most were heavily bombed, the *Me 262* squadrons were using the *Autobahns* for runways during take off and landing.

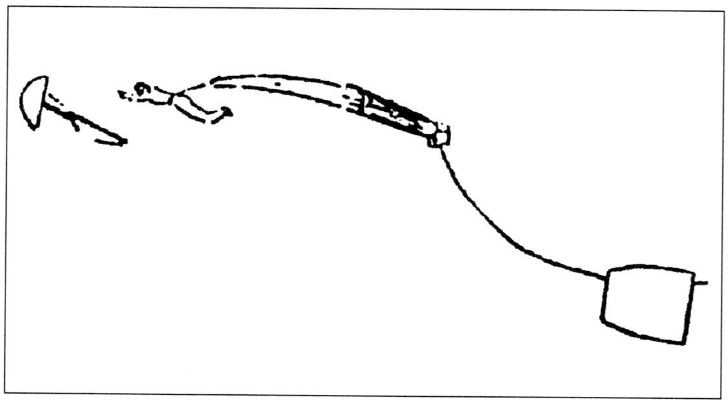

A pen and ink drawing by *DFS* showing the next stage of the *DFS 228/ DFS 346's* pilot escape system. Featured in this illustration the pilot has released himself from his couch and is being parachuted back to earth. However, the couch and plexiglass canopy are free-falling back to earth.

A pen and ink drawing from *DFS* featuring the final stage of the pilot's escape system. Here we see the pilot parachuting to earth after being ejected at a safe altitude.

MEYER - Well, we were not ready to flight test the *346* in 1945 anyway. We had thought about using Rechlin as the *346's* flight test center. Then, too, if we had defeated the Russians by this time (1945), then we would have had a lot of runway facilities again! We were thinking about placing the *346* on a bomber to be carried aloft to 10 kilometers [32,810 feet].

MYHRA - What type of bomber would have been used to carry the *346* up to altitude?

MEYER - But first we would have had to make many kinds of ground testing. We were not even ready for the ground testing program by May 1945. For example, the *HWK 509D* bi-fuel liquid rocket engines, if that was the *509* version which was being used, would have had to have been thoroughly tested on the ground and in and out of the aircraft. The next phase would have been to fly the *346* as a glider (without power), either carried by a bomber or towed to 10 kilometers altitude. The new *Arado* four turbojet-engined *Ar 234C* would have been a very good carrier aircraft. I think that we would have preferred to tow the *346* up to altitude, however. After reaching 10 kilometers the *346* would have been released and allowed to glide back to its *flugplatz*. So, we would have had to have considerable experience with the *346* on how to start and how to land it. On the third phase would have been the power flight testing, but we would not have gone to 30 kilometers altitude right away.

I believe that the *346's* initial flight testing program, that is, all the necessary learning before attempting to reach *Mach 1.0* at 30 kilometers altitude, would have taken us a full year.

We still had good aviation people by 1945, like *Willy Messerschmitt*, for example. He was a good aircraft constructor, especially his *Me 262*. When the American aviation experts inspected their first captured *Me 262*, they could not believe how light the aircraft was and the speed it was capable of, despite its relatively low amount of engine thrust. I earned my *Dipl.-Ing.* at *Messerschmitt AG. Willy Messerschmitt* frequently made fun of his engineers. For example, I remember one time *Messerschmitt* saying to one of his engineers "What is this heavy piece? Is it for a locomotive or what?" *Messerschmitt* was always stressing lightness in aircraft construction.

I am not interested in making war anymore. In the last days of the war, 25 April 1945, I lost my right arm due to a bombing attack on the City of Freliessing. And I had never even been a soldier. Only a civilian. My father died on the last day of WWI. He was getting ready to fly back to Germany. After the war I went looking for some work that was non-war related. So I became a teacher of chemistry, mathematics, and physics.

I also knew *Hanna Reitsch*. Aviation in Germany was mostly an all-male activity. Once in a while you would hear a girl's laughter, and it would be *Reitsch*. She was a great fan of *Hitler*.

End of 1st *Eberhard Meyer* Interview

A close up view of the actual 3rd cockpit cabin, built out of metal, in the USSR postwar.

The 3rd *DFS* cockpit cabin built in the USSR from drawings by *Felix Kracht* and *Eberhard Meyer*. Its entire plexiglass nose cone extended about three feet out and away from the center fuselage. This allowed the pilot to position himself on the couch prior to flight, as well as allowing him to release himself during escape from the crippled aircraft.

Eberhard MEYER
26 May 1988
Heiligenberg, West Germany

MEYER - Dr. Myhra, you are lucky that I am still here. I had a heart-attack earlier this year, and I almost died from it. It is now forty-three years ago last month that I lost my right arm during an air raid on Ainring on 26 April 1945. By the way, I have located several additional photographs of *DFS* personnel for you.

MYHRA - Oh, look, here is *Udet*!

MYHRA - You also see *Professor Georgii*, the chief of the *DFS*. These photos were taken in 1939 or 1940 at *DFS*-Darmstadt, just before or shortly after the beginning of the war. *Udet* came often to *DFS*-Darmstadt, and *Georgii* is showing him some of our project work. I am also in the photo, but on the far left. See here, this man is *Felix Kracht*. He always looked like this...a man with a hat and his hands in his pants pockets. The tall man near the center is *Hans Jacobs*. Later our whole operation was moved east to the city of Braunschweig, near Hannover, to protect *DFS'* activities from war damage.

A poor quality photo showing the pilot positioning himself in *DFS'* 3rd cockpit for the *228* and *346*. This particular photo comes from the USSR.

MYHRA - When I first visited with you on 29 May 1987, you mentioned that the French had obtained a complete copy of the *DFS 346's* drawings. How did they obtain these drawings?

MEYER - Well, the Americans arrived first at *DFS*-Ainring, and then a short-time later at *Siebel*-Halle. It was either at Ainring or Halle that they found and then took a copy of all the *DFS 346* drawings. French military intelligence people came later, too, plus others. Once the Allies—for example, the Americans, English, and French—found out that we Germans were really seeking to break the speed-of-sound with the *DFS 346*, well, then each of them wanted this material, too. They all wanted to fly beyond the speed-of-sound. By the way, if you see any of the *DFS 346* documents you will see *Kracht's* name on them, for as chief of the *DFS 346* project, he was the one who signed them. The French obtained a copy of the *DFS 346* drawings, too, about the same time as the Americans, but the French did not have the aviation specialists able to handle/construct a project like the *DFS 346*. So, the French invited *Georgii*, *Kracht*, *Friedrichs*, *Nestler*, *Niese*, and *Mödel* to France to continue working on the *DFS 346*. These men all went voluntarily to France. The research institute where they were located was near Paris. But I did not want to work on military aircraft anymore. I did not want to participate in war-making machinery, so I didn't go with them. In mid 1946, I was invited to come to Paris, too, by *Kracht*, but by then I had a good job, and so I did not need aviation work anymore. But life in Paris was good for my former colleagues. They all had plenty of food to eat and good housing. *Kracht* continued to be their leader. They all worked there together for about two years...from about August 1945, and up to about mid 1947...and then the French decided that their work on a supersonic

A view of the open, empty 3rd *346* prototype pilot's couch looking back to its rear bulkhead.

R&D aircraft was simply too expensive and that they could not afford it anymore. So, the French government halted all this work. Then, I believe, the French government handed over all their documents of the *DFS 346* and related research work to the American government. The Americans, on the other hand, had many, many aviation specialists and money, so they did not need people like *Felix Kracht* and the others. All the Americans needed were the design calculations upon which the *DFS 346* was based, and then they could build their own *346*. This they did, by the way, in the form of the *Bell X-1*. That is why they did not need to take the *DFS 346* at Halle, because it was only partially complete anyway.

MYHRA - Who in the French government was in charge of the *DFS 346* research project?

MEYER - I don't know, because I did not go to France, nor did I talk to my former colleagues even after they returned to Germany, with the exception of *Niese*. He later became a teacher, like me, here in Germany. Better to ask *Kracht*, *Niese*, or *Mödel*. I taught in a "*Reuter-School*," that is, a school which is known as an "unstructured" school. These were schools which had no set curriculum, but the students are pretty much free to study whatever interests them. This type of instruction was not allowed during the *Hitler* years of 1933-1945. But right after the war we went about establishing them again. I taught physics and astronomy. Did you know that after the war citizens of Holland sent food to us Germans? It was sort of a "care" package, despite the fact that our government had made war upon Holland. Also, those German families which had no homes left could send their children to foster homes in Norway and Switzerland.

MYHRA - Did *Alexander Lippisch* have anything to do with the *DFS 228* or *DFS 346* projects?

MEYER - No, nothing to do at all with the *DFS 346*, its general planform, nor the selection of *Siebel*-Halle as the constructor. Nothing on the *DFS 228*, either. *Lippisch* was mainly a sailplane specialist at *DFS*, although he was the father of the *Me 163* rocket fighter.

MYHRA - The reason I ask is that I have been told that *Lippisch* did sit in on some of the early brainstorming discussions on the *DFS 346*. What input did *Professor Quick* of the *DVL* have on the *DFS 346?*

MEYER - Yes, *Lippisch* had been on the *DFS 228's* and *DFS 346's* mailing list. *Professor Quick* was a wind tunnel specialist at *DVL*-Berlin/Adlershof. They had a small supersonic wind-tunnel there. *Quick* was working closely with us on the aerodynamics of both *DFS* machines. He had built an aluminum scale model of both for use in the wind tunnel. The biggest supersonic wind tunnel in Germany at that time was in Braunschweig, but the tunnel at Berlin could also reach supersonic speeds.

MYHRA - How large a wind tunnel model of the *DFS 228* and *346* did *Professor Quick* construct? Was it one meter or more in length and wing span? Or was it smaller?

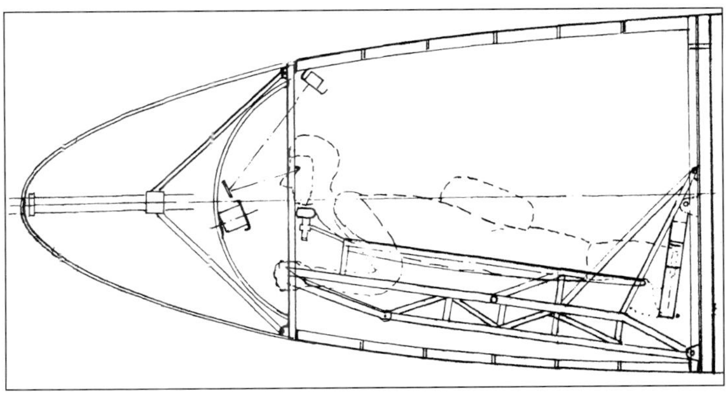

A pen and ink simple drawing *DFS'* 3rd *228/346* prototype cockpit cabin from *DFS* featuring the cabin's port side.

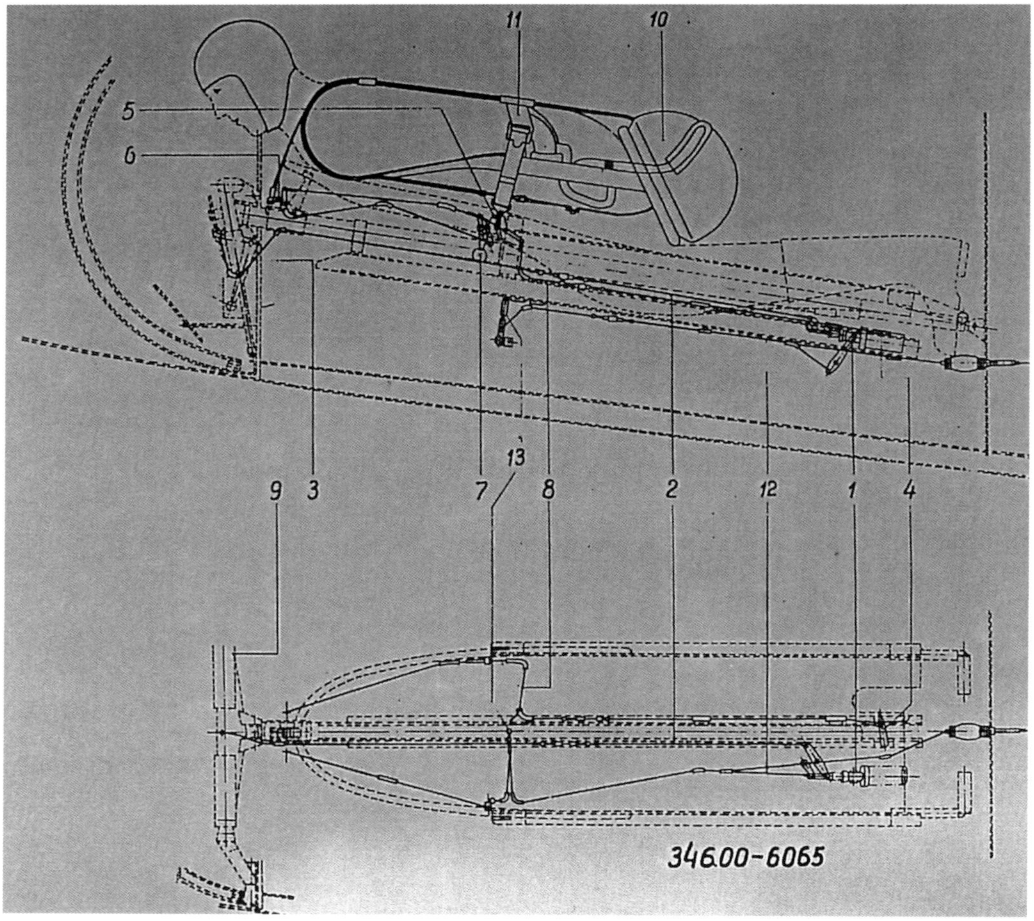

A pen and ink simplified drawing from the USSR post war of *DFS'* 3rd *228/346* prototype cockpit cabin. The major change from the 2nd and 3rd prototypes are the automatic pilot ejection mechanism.

01 - ejection control cylinder
02 - parachute locking device cord
03 - parachute manual locking lever
04 - automatic fastening belts release lever
05 - side locks of pilot's fastening belts
06 - front lock of pilot's fastening belts
07 - manual release lever
08 - cord to the pilot's fastening belts lock
09 - parachute
10 - pilot's parachute
11 - pilot's fastening belts
12 - pilot's bed lock releasing (manual or automatic) cord
13 - bed lock

Below: A pen and ink drawing featuring the explosive device on the 3rd *228/346* to propel the pilot's couch/bed and plexiglass cockpit canopy out of the cockpit cabin.

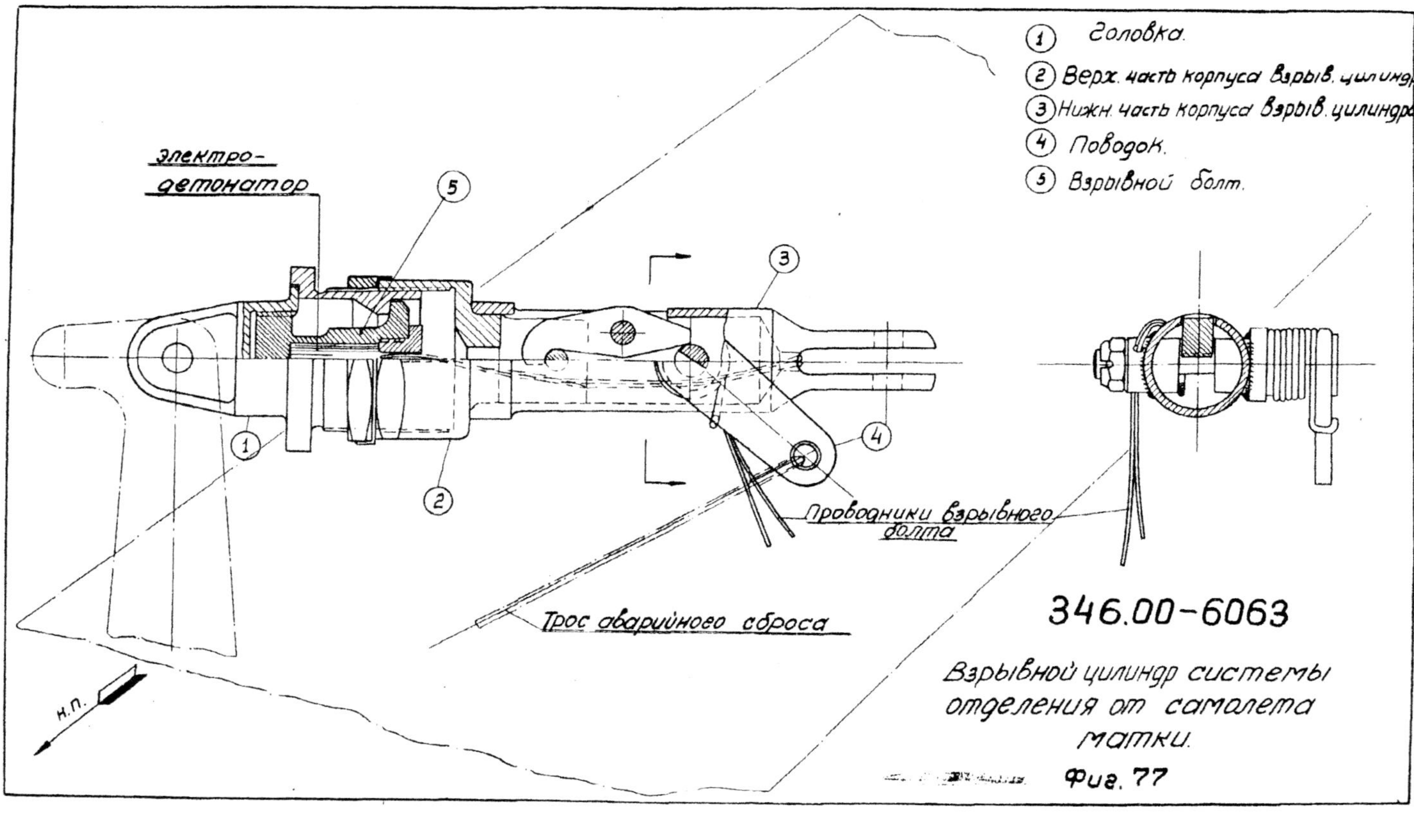

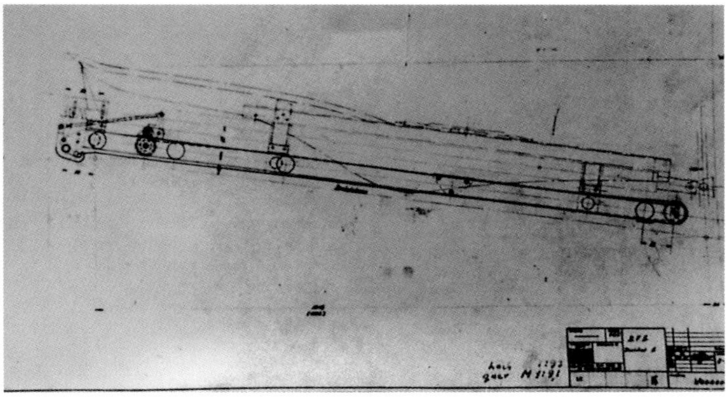

A poor quality pen and ink drawing from the USSR featuring the 3rd *DFS 228/346's* couch/bed's steel wheel and rail mechanism which allowed the pilot and couch to easily roll out of the cabin after being propelled by the *Karl Bütter* designed explosive bolt system.

MEYER - No, not that large. It was a very small model, because Berlin/Adlershof had only a very small wind tunnel.

MYHRA - How much time did *Professor Quick* spend testing the *DFS 228* and *346's* planform in *DVL's* Berlin wind-tunnel?

MEYER - A lot of time. *Quick* tested both profiles...their fuselage, and with the *DFS 346* with different wing profiles and shapes at Berlin/Adlershof. *Quick* gave the *DFS 346* a lot of attention. Different wing thickness profiles were tested, some with sweep back and some with a straight wing. *Quick* had electrical wires attached to this model so he was able to measure the different stresses associated with swept back and straight wings. I recall that three or four different sets of wings with different degrees of sweep back were tested in the wind tunnel. We found that the best sweep back for our purposes was between 40° and 50°, with 45° about the best. In addition, we tried several different tail configurations that were built and wind tunnel tested, too. All in all, I think that we had up to twenty-four different wind tunnel combinations of wings and tail units for testing.

MYHRA - Did these four separate wings have roughly the same square meters of surface area, about the same aspect ratios, and so on?

MEYER - Yes, they were different in terms of profile, thickness, sweep back, and so on.

MYHRA - The final *DFS 346* planform had a wing profile of 12 degrees. Had other wings with different thickness' been tried in the *DVL* wind tunnel, too? People say that the *DFS 346* with a 12° wing thickness would not make *Uberschall.*

MEYER - I do not recall if we tried other wing thickness. Anyway, 12° was about as thin as we could make the wing at that time based on the construction materials we had available to us, plus we did not know then how strong the air flow was going to be over the wing at the speed of sound. The 12° wing thickness was selected for strength, too, if we were intending to fly at 30 kilometers altitude [98,400 feet]. This is why we had the *346* fitted with all kinds of electric stress measuring points.

MYHRA - Sure. People talked about the "sound barrier" and the so-called "wall in the sky." In what year did *Professor Quick* make all these wind tunnels tests on the *DFS 228* and *DFS 346?*

MEYER - During 1943. I did not have too much to do with these tests, and I believe that *Professor Quick* was able to reach the speed of sound with the final version of the *DFS 346* in *DVL's*-Berlin/Adlershof wind tunnel.

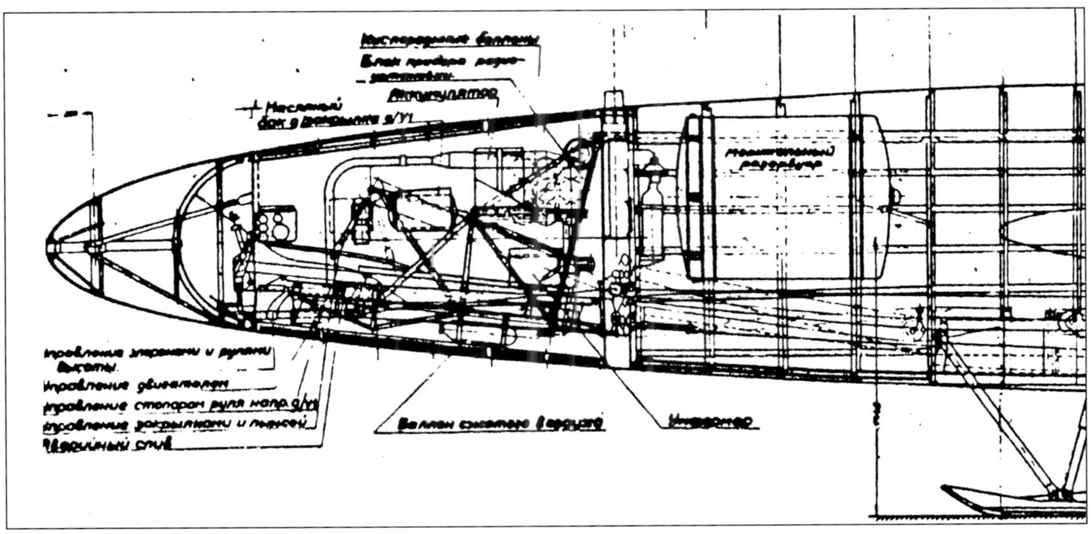

A highly detailed pen and ink drawing from the USSR post war of *DFS'* 3rd prototype pilot's couch. Featured is the prototype's port side. Beginning from the top left (starboard side) side of the drawing and down:
- oxygen tanks - battery
- radio unit - oil tank
Beginning from the port side nose of the drawing and down:
- aileron and elevator control
- *HWK 509* engine controls
- rudder lock control
- flaps and metal landing skid control
- emergency drain control
Mid lower bottom of photo:
- pressurized air tank
- unknown

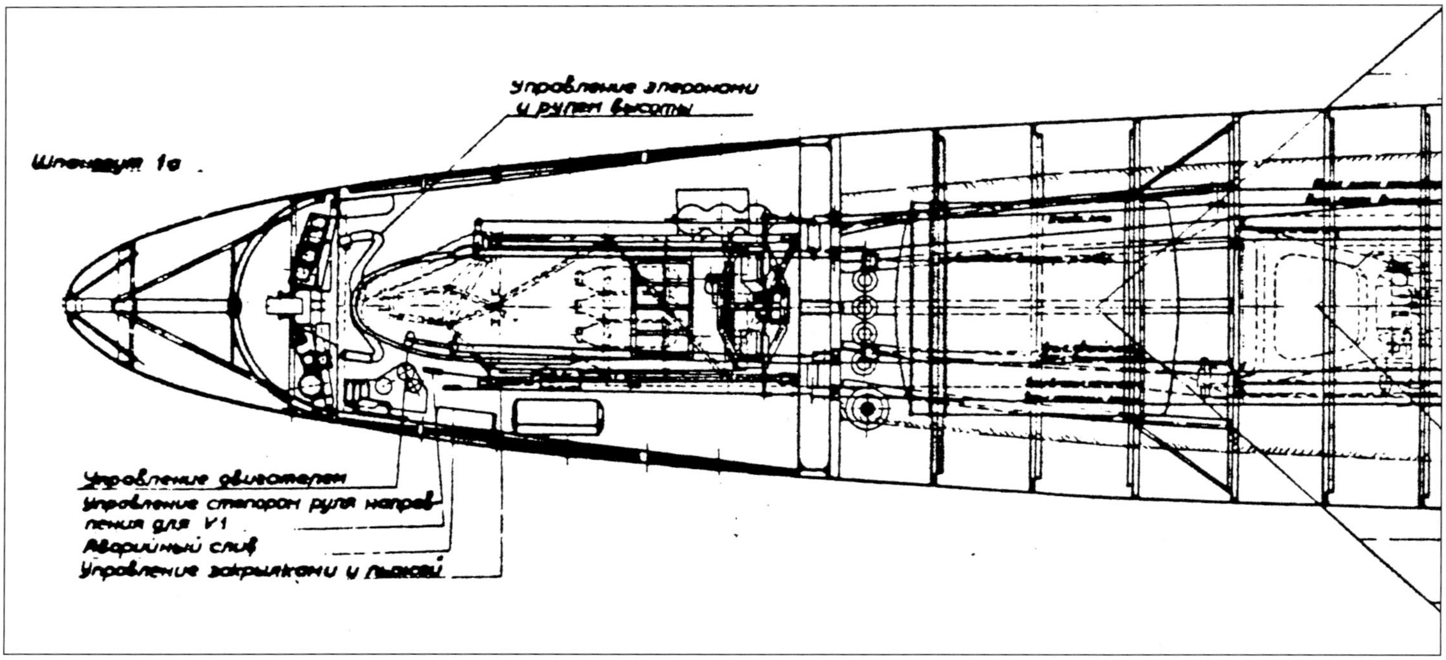

A highly detailed pen and ink drawing from the USSR post war of *DFS'* 3rd prototype pilot's couch as seen from above looking down. Top to bottom:
- aileron and elevator control
- frame #1a
- engine control
- rudder lock control
- emergency drain control
- flap and metal landing skid control

MYHRA - Were people such as *Kracht* and *Georgii* happy with the *228* and *346's* general final planform based on the wind tunnel testing?

MEYER - Yes, I recall that the wind tunnel test results showed that we had made a good design for the *DFS 228* and *DFS 346.* It appeared to us that the *DFS 346* would perform well at supersonic speeds. We would have tried different wing profiles during its flight testing program. However, the scale model tests in *DVL's* Berlin/ Adlershof wind tunnel turned out very positive. The final wind tunnel model showed no undesirable flight characteristics.

MYHRA - Was the *DFS 346* wind tunnel model tested with a forward sweeping wing?

MEYER - No. We in *DFS* were not interested in the forward sweeping wing at this time. We were really into sailplanes, and in late 1943-1944, sailplane transports and the problems of aircraft approaching supersonic speeds. We had designed the sailplanes used in the capture of the Belgian *Fort Eben Emeal* in May 1940. *Udet* had made the plans for this operation after *Hilter* ordered him to capture the *Fort.* Eleven gliders carrying a total of seventy-eight glider-men landed on what was then called the world's most "impregnable" fort, and within twenty minutes neutralized its 780 man garrison. The next day the fort capitulated. With this feat, glider warfare was born. I believe that these men were among the first soldiers of what was to become *WWII* to receive the *Ritterkruz.*

MYHRA - Did *DFS* at any time build a wooden mockup of the *DFS 346?* Did *DFS* ever want to build a *DFS 346* sailplane and test its configuration like you did with the *DFS 228?*

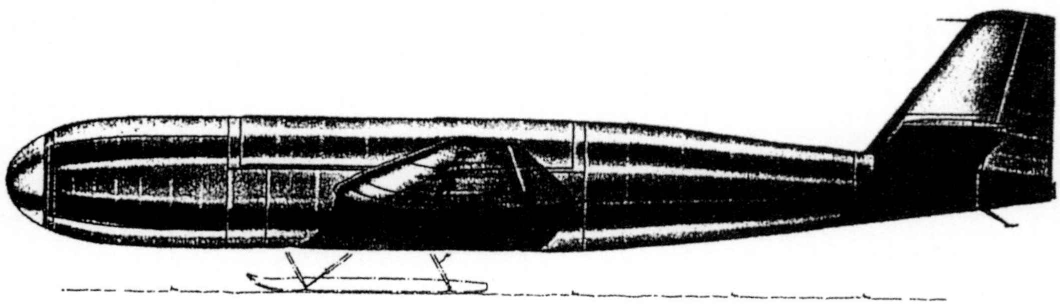

A pen and ink drawing of the *DFS 228*, incorporating its 3rd prototype cockpit cabin featuring the port side. This prototype featured a full plexiglass nose unlike the round port hole-like window in the *DFS 228 V1*, the squarish window in the *DFS 228 V2.*

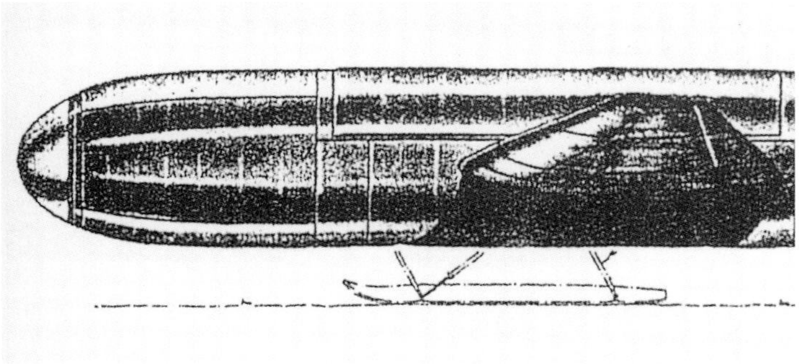

A pen and ink drawing featuring a close-up of the 3rd *DFS 228/346* prototype cockpit cabin as mounted on the *DFS 228*.

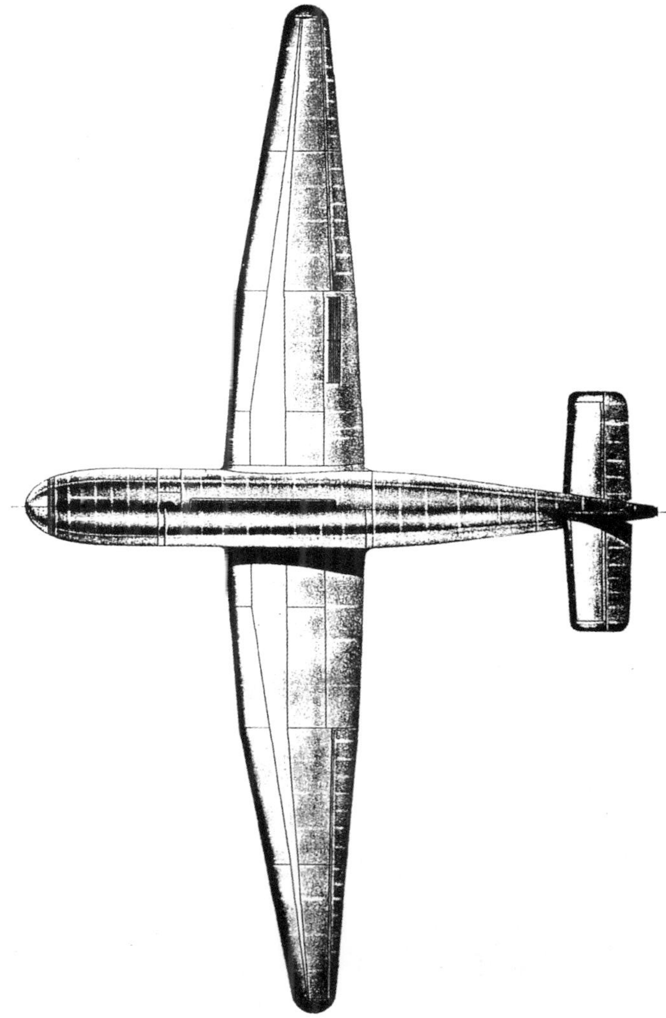

A pen and ink general arrangement drawing from *DFS* of their *DFS 228 V3* equipped with *DFS'* 3rd prototype cockpit.

MEYER - We did not build a wooden mockup of the *346*. But we did build a new cockpit cabin for the *DSF 228*, and it would have been used on our *DFS 346,* too. They were all built by *Mödel*, who also built the several cockpit cabins for the *DFS 228* out of wood. But the fuselage, wings, and tail assembly of the *346* were to have been built out of aluminum. No, we did not have any interest or time for building a *DFS 346* prototype. We had no experience in aluminum.

MYHRA - How many *DFS 346s* was *Siebel* told to construct at Halle?

MEYER - I recall that constructing the *DFS 346* was getting to be an expensive activity for the *RLM* at that time in late 1944. I believe that *Siebel* was told by the *RLM* to build only two prototypes and one fuselage-frame for stress testing.

MYHRA - At any time was a retractable landing gear considered for the *DFS 346?*

MEYER - No wheels, because the landing speed was to be very fast. Only a landing skid like on our *DFS 228* and the *Me 163* was to be used for the *DFS 346*.

MYHRA - Who was to have been the *346's* test pilot? Anyone from *DFS*?

MEYER - No, I dc not think anyone from *DFS*. Someone from *Siebel*, I suppose, but I don't know who. The *DFS 346* would have never been ready to fly in wartime Germany.

MYHRA - Did *Professor Eugen Sänger* have anything to do with the *DFS 346* project?

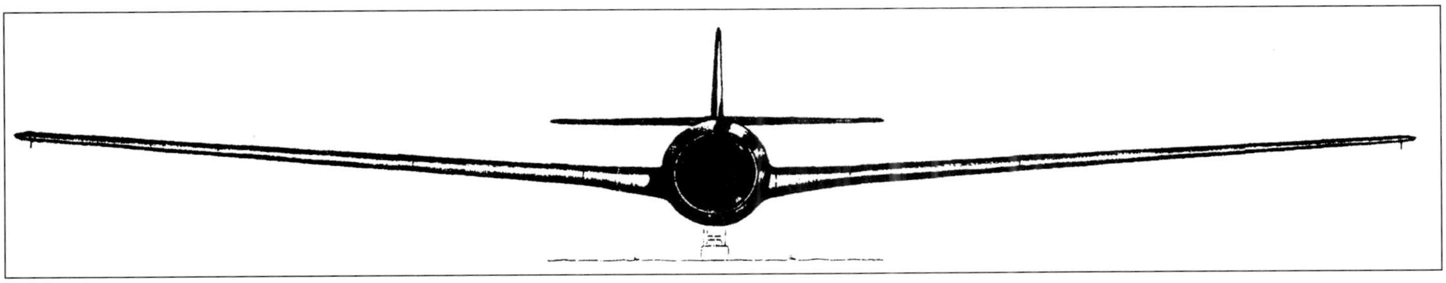

A pen and ink drawing based on *DFS'* 3rd prototype cockpit cabin as viewed standing in front of the *HWK 509* bi-fuel liquid powered sailplane *DFS 228 V3*.

A scale wind tunnel model of the *Junkers Ju 388*. It has been said that *DFS*-Ainring was considering replacing their *Do 217 K3* carrier aircraft with the newer *Ju 388*. It did not happen, and *DFS* continued employing the *Do 217 K3* right up to war's end.

MEYER - No connection. *Sänger* was working on other things at *DFS*, such as his so-called orbital bomber and researching the ram jet engine.

MYHRA - You mentioned earlier that *DFS*-Ainring had or wanted to try a different nose cone shape on the *DFS 346*. What do you mean?

MEYER - We were considering several highly aerodynamic shapes for the *DFS 346's* nose cone. We did not consider placing the pilot in an upright position. We only planned on placing the pilot on a couch, so that he might not "black out." We had done a lot of testing with a pilot laying on a couch with the *DFS 228*. Some initial testing was done by the students at the *Berlin Akaflieg*, first with a sailplane, and then with a twin motor monoplane designated the *Berlin B-9*.

MYHRA - Why would the *DFS 346's* pilot suffer "black out" in attempting supersonic speeds, therefore needing some protection from it? I can understand if the pilot was making all sorts of aerobatic maneuvers, tight turns, dives, and so on, then "black out" might be a problem. But why would "black out" be a problem in the *DFS 346*, which would have been flown in a straight line with gentle turns and so forth?

A direct front on view of the *Arado Ar 234C*. It would have had the power and carrying capacity to serve as a launch platform for the *DFS 228* as well as the *DFS 346*. Scale model and photographed by *Günter Sengfelder*.

***DFS*-Ainring was also considering replacing their aging *Do 217 K3* with the four turbojet-powered *Arado Ar 234C* as seen here from its nose port side. Scale model and photographed by *Günter Sengfelder*.**

MEYER- Yes, of course. We did not build the *DFS 346* to make loops, dives, and so on.

MYHRA - I recall that *Walter Horten* once told me that "black outs" happened more often while one was making high speed turns. It was then the blood drained from the brain and the pilot "blacked out." Black outs did not happen so much during maneuvers or dog fighting.

MEYER - Sure. But we were thinking "what if the power of the *HWK509's* dual combustion chambers were so great, or if the pilot made a mistake and "black out" conditions developed? We wanted merely to protect the pilot, nothing more, through the use of this couch. That was all. The lack of a standard cockpit canopy was not the reason for the pilot's couch. Although a pilot laying on his stomach presented a clean, smooth fuselage, this was not the reason for the couch. Protecting the pilot from "black out" was our main concern. We really did not know what might happen to the aircraft and/or pilot at that moment when the speed of sound was broken.

MYHRA - I have a question about the *DFS 346's Walter HWK 509* rocket engine. We know that the *346* had two exhaust nozzles. Could the *346's* rocket engine have been similar to the one used in this *Ju 248/Me 263?* I think that you mentioned the *Ju 248* last year when I was here. You told me that you went to *Messerschmitt AG*-Augsburg from time to time to consult on the *Ju 248*.

MEYER - I don't remember anymore the type or number of *Walter* rocket engines used on the *DFS 346*. I went to *Messerschmitt AG*-Augsburg to visit with *Voigt* on the problems dealing with the *Me 163*, especially in recovering from high speed dives. Sometimes a *Me 163* could reach up to 1063 km/h [660 mph] when making passes at *B-17s*. So the *Me 163* was flying to the edge of the speed of sound.

MYHRA - Would you identify the people in these several photographs?

MEYER - The two pictures with the mountains in the background are in the Alps mountains where we tested several different types of gliders. We took them up the mountain, and then they were flown off the mountain so that in the process we could test the glider's high-pressure cockpit cabin at altitude.

MYHRA - And the people around the table?

MEYER - The photo of the people gathered around the table were taken at *DFS-Ainring*.

End of 2nd *Eberhard Meyer* Interview

Werner NESTLER
2 May 1988
Ostfildern (Stuttgart), West Germany

NESTLER - I was responsible for the *HWK 509* rocket engine installation on the *DFS 228* at *DFS*. I joined *DFS*-Ainring in 1942. I was given the job of engine specialist/expert because I liked the work, and also because no one else knew much about them or even wanted to learn. I had had some experience in turbojet engines prior to coming to *DFS*. *HWK* had its main factory/offices in Kiel. *HWK* also had an engine production facility in the Tyrol mountain area (Alps) in Austria.

I remember that about July 1945, *Professor Georgii*, former head of *DFS*, gave the American Intelligence people in Germany a full set of plans for the *DFS 346*. Getting back to the *228's HWK 509D*, it was bench tested at the *DFS*-Hörsching's airdrome. Hörsching was considered *DFS'* personal airport, and this is where *DFS* tested all their project aircraft, such as the *228*. The *228* was never flown with rocket power, although one had been installed in the civilian registered *D-IBFQ*, the *DFS 228 V1* prototype. A second *DFS 228* prototype (*V2*) had been destroyed during flight testing, at *Eprobungsstelle*-Rechlin. The *D-IBFQ* survived the war and was taken by the American occupational forces to Stuttgart. *DFS* had troubles with the *HWK 509D* engine even on the test-stand,

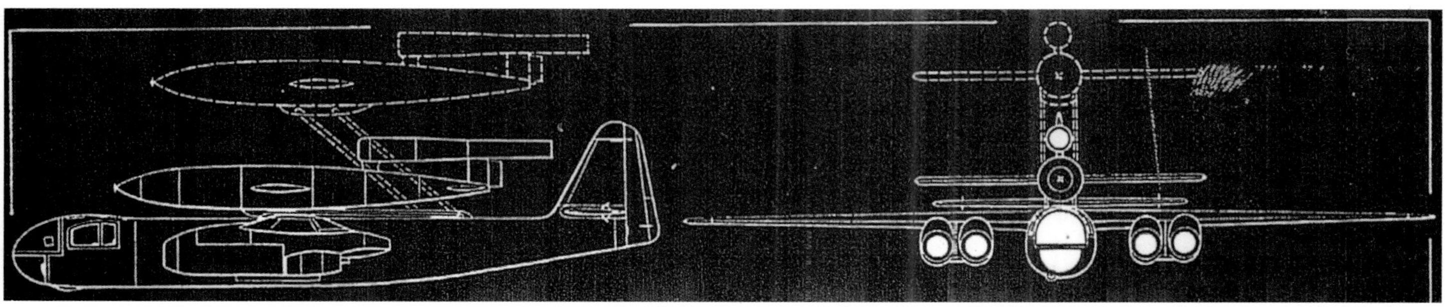

A pen and ink drawing from *Arado* featuring how their *Ar 234C* would be customized to carry the *Fi 103* flying bomb. Hydraulic articulated arms would raise the *Fi 103* high enough to clear its tail assembly and then released. The *DFS 228* and *346* could be launched the same way.

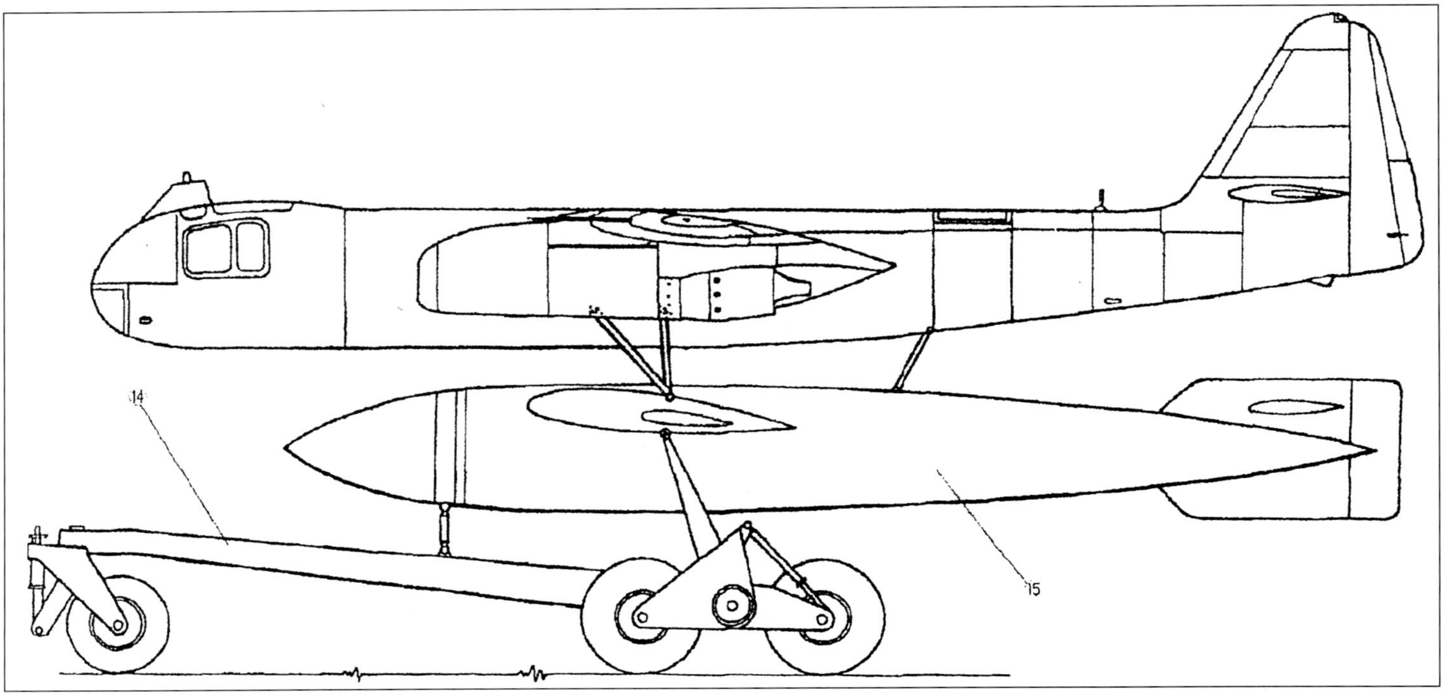

The *DFS 228* or *346* could be carried beneath the *Ar 234C*. A three-wheeled dolly would have to be used for take off as shown in this pen and ink drawing by *Arado*.

due to the "off/on" metering system required for the *DFS 228's* "*Sagezahn*," or sawtooth reconnaissance flight pattern. The fuel metering and fuel line pressurizing problems on the *HWK 509Ds* were not corrected by the time of the surrender in May 1945, either.

Dr.-Ing. Eugen Sänger tested his ram-jet engines at Hörsching, too, but he did not appear to have anything to do with the DFS 228 or the *346* projects. The pulse-jet powered *Me 328* was also flight tested by *DFS*-Ainring test pilot *Rudolf Ziegler* at Hörsching. *Ziegler* was also the *DFS 228's* test pilot, and he flew the *228* up to forty or more times. But each time only as a sailplane with its *HWK 509D* rocket engine not operating.

I do not remember much about the *DFS 346's HWK 509C* rocket engine with its dual combustion chambers. Nor do I recall whether the *509C* was the engine of choice for the *DFS 346's* designers and builders. The *HWK 509C* was the engine specified in the technical drawings passed on to *Siebel*-Halle. However, I do not know if *Siebel* officials changed the single *509D* to two *509Bs* or some other customized *509* unit. Two HWK *509B-1s* were not placed in the aircraft, although the *346* was designed to hold two standard *HWK 509B*-1 rocket engines, one over the other. I do not know the mounting arrangement or how far forward in the *346's* center section they were located. *DFS'* only responsibility on the *346* was to make the construction drawings and plans and then give them to *Siebel* officials. In addition, *Siebel* was totally responsible for locating and obtaining all the necessary equipment, parts, supplies, and engines needed to build the *346*. *DFS* had no responsibility in this whatsoever, but had only to be available if there were questions from *Siebel*-Halle people on the plans and drawings. *DFS* supplied advice if needed by *Siebel* people, and then only from time to time. After *DFS* delivered the drawings to *Siebel* in November 1944, *DFS*-Ainring people moved on to other *RLM*-assigned aircraft projects. I never got to *Siebel*-Halle regarding the *DFS 346*, and I assume that *Siebel* had assigned someone from their factory as the chief *HWK 509* rocket engine specialist on the *DFS 346*.

End of *Werner Nestler* Interview

Hermann ZITTER
5 November 1986
Bad Homburg, West Germany

MYHRA - The normal ceiling of a *Dornier Do 217* was about 29,527 feet (9 kilometers). How was it possible to extend its ceiling to 39,370 feet (12 kilometers) in order to carry your *DFS 228* up to its starting altitude?

ZITTER - The *Do 217K* that we used to tow and/or carry the *DFS 228* had two special turbo-charged engines...*BMW 801TJs*. I don't recall anymore what changes were made to these engines, but they were powerful high-altitude types.

MYHRA - Was any other aircraft used for towing, such as the *Heinkel He 177*?

The *DFS 228* could have been carried to altitude under the wing of the *Ar 234C*. Shown in this photograph from the USSR is the *DFS 346* being carried up to altitude under the starboard wing of an interned *B-29* heavy bomber.

ZITTER - Yes, but not with us, because we were getting a *Junkers Ju 388* in the future to replace our old *Do 217K*.

MYHRA - Tell me about the test-flight results of the *DFS 228*. I have heard that its general handling was satisfactory. Some reports refer to the *228* as the "*Lippisch* Glider."

ZITTER - Whoever called the *DFS 228* the "*Lippisch Glider*" is spreading misinformation. Its overall configuration...in fact, the balance of its flight controls, well, they are always the problems in most new aircraft. In addition, flight controls, or their balance, are problems in the stratosphere, and they are really problems when you break *Mach 1.0*. This was our experience on the *DFS 228*. You see, our *228* was a *HWK 509D* powered experimental aircraft, and you expect weird things with experimental aircraft. Do you know that *Kracht* and others were considering replacing the *HWK 509D* with the new *BMW 3390A* bi-fuel liquid rocket engine? It would have been a better power plant for the *DFS 228*. As I recall, the *BMW 3390A* had an electrical starter, plus its own electrical generator. We had been talking about installing a *3390A* in one of our *DFS 228s* as early as January 1943. It would have been powerful. This rocket motor had initially been intended for the *Me 163B*. There was this engineer at *BMW*-Spandau by the name of *Helmuth Zborowski*. He believed that with his *3390A* rocket motor the critical velocity of the *DFS 228* could be surpassed in climbing! In fact, we had discussions with *Zborowski* regarding the installation of the *BMW 3390A* in one our *228s*. *Zborowski* claimed that fuel consumption would be much lower than the *HWK 509D*. Plus, he was telling us that the combustion chamber could be separated from the motor head without any problems so as to improve the *DFS 228's* center of gravity. For some reason, probably ventilation, I recall that the *BMW 3390A* would have needed an air scoop on the *228's* fuselage. I do not know why our *228* would have been referred to as something connected to *Lippisch*. As far as I know, he did not have much, if anything, to do with its design. He was into tailless and delta designs at that time.

MYHRA - Yes. However, the *BMW 3390A* was still an experimental rocket engine, and as far as I know, it was never tested in a flying machine. I am not sure that it was even ready for field testing. What is the fastest forward speed the *DFS 228* ever flew as a sailplane?

ZITTER - I really don't know, but anyway, its highest forward speed would have come as a glider and probably in a dive.

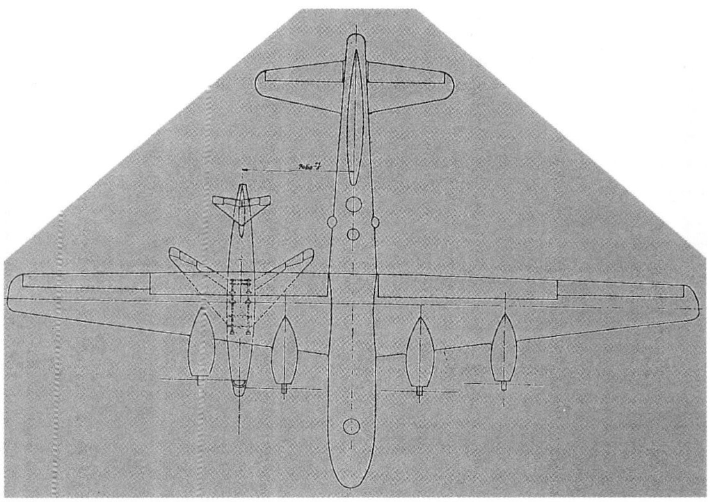

A pen and ink drawing from the USSR featuring the general arrangement of a *B-29* carrying a *DFS 346* under its starboard wing. The *DFS 228* could have been carried aloft the same way, however, by using the *Ar 234C*.

An actual photograph of an American *B-29* in the USSR post war with a *DFS 346* hanging from its starboard wing...just the same way a *DFS 228* would have been carried by a *Ar 234C*.

MYHRA - I have read, for example, that the *DFS 228* was supposed to have an anticipated sea level speed of 560 mph [901 km/h].

ZITTER - Well, that speed of 901 km/h was only theoretical, since we never made any powered flights.

MYHRA - Was the *DFS 228* intended to reach *Mach 1.0* in level flight?

ZITTER - Oh no, only subsonic and well under *Mach 1.0*. However, the *DFS 346* was supposed to fly two times the speed of sound.

MYHRA - Was any pilot-protecting armor to be fitted to the *DFS 228?*

ZITTER - I don't know for sure, but I do not think so.

MYHRA - Why did the *DFS 228* have the civil registration number *IBFQ*? It seems very unusual to have a research aircraft with a civil registration number.

ZITTER - Yes, it had a civil registration number because our *DFS 228* was intended for civilian experiments at the *Rhön/Wasserkuppe*, so a pure military research machine showing up at the *Wasserkuppe* would not have looked so good. You should ask this question of *Eric Klöckner*. By the way, *Klöckner* holds a record for the highest altitude in a sailplane. He reached 36,000 feet [11 kilometers] once in a sailplane at the *Rhön/Wasserkuppe*.

MYHRA - I have read that the idea for the *DFS 228* was first presented in a paper before the Society of German Aeronautical Research in November 1944. Do you know who this person was?

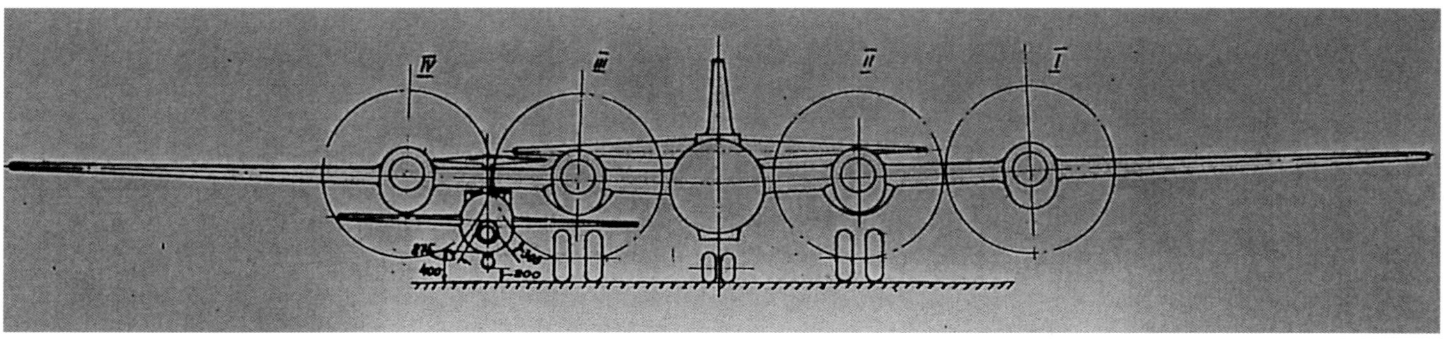

A pen and ink drawing from the USSR featuring a nose on view of a *B-29* carrying the *DFS 346* under its starboard wing. An *Ar 234C* would have served the same purpose as the *B-29*.

The *DFS 228 V2* as seen from its nose port side. Notice that the *228 V2*, as seen through the cockpit canopy, carried its pilot in the prone position laying on a couch. Digital image by *Mario Merino*.

ZITTER - No. Perhaps *Felix Kracht* or Professor Georgii would know all these details. I do not know why these two aircraft...the *228* and *346* were even built in our Institute in the first place. Because these two aircraft created many, many problems with their construction, the need to test the *Walter HWK 509D* bi-fuel liquid rocket engine, over and over again, and so on.

MYHRA - Who was the engineer in charge of the *HWK 509D* rocket engines at *DFS*?

ZITTER - I do not know. I think *Helmut Walter* himself, because he and several members of his engineering team came frequently to *DFS* to give assistance on these rocket engines.

MYHRA - The operational plan was to tow the *DFS 228* up to 32,800 feet [10 kilometers] and its *HWK 509D* rocket engine would be ignited. Rocket power would then propel the aircraft up to 75,440 feet [23 kilometers]. Is this correct?

ZITTER - Yes.

MYHRA - I have read that there were two lines of development planned for the *DFS 228*: increase its maximum speed and altitude; and second, to improve its aerodynamic lines to increase efficiency.

ZITTER - Yes, this is correct.

A nose on view of the *DFS 228 V2* featuring its pilot in the prone position. Digital image by *Mario Merino*.

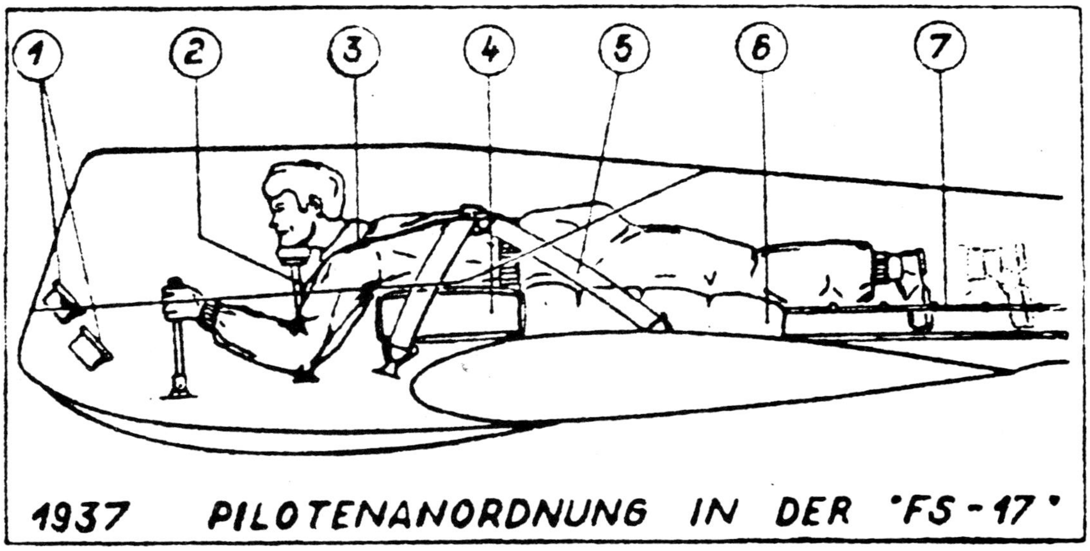

Piloting aircraft in the prone position, that is, while laying on a couch had been tried in Germany in the 1930s. A pen and ink drawing of the *Flugtechnische Fachgruppe-*Stuttgart *FS 17* was one such example which *Felix Kracht* and his assistant *Eberhard Meyer* drew upon when they designed the prone flying position in their *DFS 228 V2.*

MYHRA - Also, I understand that the *DFS 228* was to have been used for short range, high altitude photo reconnaissance? What types of cameras would have been used? What about infrared cameras?

ZITTER - Yes to both items. Really the *DFS 228* was the forerunner to your American *U-2*. Our *DFS 228* was to have had a special camera for this purpose designed by the *Zeiss Werk*. I know for a fact that the *Zeiss Werk* was working on a camera especially for the *DFS 228*. I recall that they were doing a lot of research leading up to a high altitude camera.

MYHRA - I have heard that up to twenty-five *DFS 228s* were to be built?

ZITTER - This I do not know. But the *228's* fuselage aerodynamics was still being investigated in our own wind tunnels at *DFS*-Darmstadt as late as March 1944. These wind tunnel tests included angles of attack from minus 3° to plus 6° at sideslip, and angles ranging from 0% to 9% and at constant aerodynamic pressure.

MYHRA - *Horst Götz* told me once that *Oberst Knemeyer* was pushing for rocket powered reconnaissance aircraft because the British had their own jet powered fighters coming into the war in late 1944, and certainly by early 1945. So, too, with the Americans and their turbojet-powered *P-80*. Both of these aircraft would have been flying in Europe by the Summer of 1945, so both the *Arado Ar 234B* and *234C* may have not have been safe any longer as they were with the *P-51*s and *Spitfires*.

Felix Kracht's "Rheinland"* sailplane as seen from its port side. This sailplane featured the prone piloting position *Kracht* favored in his *DFS 228 V2.

The *"Rheinland"* with its cockpit canopy removed. Notice that the pilot put his chest up against the vertical portion of the support board and his chin on the horizontal portion of the board.

The *"Rheinland"* by *Felix Kracht* appears almost ready to be towed up to altitude. The pilot can be seen prone in the cockpit.

ZITTER - That is possible. I recall that they had a great deal of difficulty with the *Ar 234B's* narrow width landing gear. Its fuselage was very narrow, too, and the aft fuselage was the only place where they could store gear and photo reconnaissance cameras.

MYHRA - Why the "swordfish" decal put on the fuselage nose of the *DFS 228?*

ZITTER - *Eugen Sänger* was the designer of the "sawtooth" concept, that is, power up, glide down, then power up again like the swordfish's bill, and this is why that fish was on the nose of the *DFS 228. Sänger* called the *DFS 228's* flight path the "*Sagezahn*," or sawtooth, just like the swordfish's tooth filled bill.

MYHRA - So the main purpose of the *DFS 228* would have been strictly high altitude photo reconnaissance?

ZITTER - Well, no, not at first. *Kracht* always viewed the *DFS 228* as a study to determine the most effective way of releasing a pilot at extreme altitudes without injury to themselves. It was the pressurized cockpit cabin that *Kracht* was really interested in. He really didn't give a damn about photo reconnaissance, but he needed an excuse for putting all that design effort and construction expense into it. So, I guess *Kracht* told those people at the *RLM* what they wanted to hear. *Kracht*, you see, was at the time still an avid sailplane enthusiast at heart, and he never forgot what had happened to his colleagues *Werner Blech* and *Heinz Scheidhauer* in their *Horten 3* all-wing sailplanes at the 1938 *Rhön/Wasserkuppe Summer Com-*

An example of the prone flying position sailplane *"Rheinland"* by *Felix Kracht* on display in pre *WW2* Germany.

Entwurf:	FFG Stuttgart
Hersteller:	FFG Stuttgart
Baumuster:	FS 17
Bauform:	Tiefdecker
Bauart:	freitragend
Verwendungszweck::	Versuch
Besondere Merkmale:	liegender Flugzeugführer

Abmessungen:

Spannweite	10,00 m
Länge, gr.	5,10 m
Höhe, gr.	1,82 m
Rumpfbreite' gr.	0,58 m
Rumpfhöhe, gr.	0,73 m
Rumpfquerschnitt, gr.	0,35 m²

Flächeninhalte:

Tragfläche mit Querruder	12,00 m²
Querruder, ges.	1,22 m²
Landeklappen, ges.	0 m²
Bremsklappen, ges.	0 m²
Höhenflosse	0,00 m²
Höhenruder	1,80 m²
Höhenleitwerk	1,80 m²
Seitenflosse	0,00 m²
Seitenruder	0,90 m²
Seitenleitwerk	0,90 m²

Gewichte:

Flugwerk	80 kg
ständige Ausrüstung	4 kg
Leergewicht	84 kg
zusätzliche Ausrüstung	6 kg
Rüstgewicht	90 kg
Zuladung	80 kg
Fluggewicht	170 kg
Zuladung/Rüstgewicht	0,889

Tragfläche:

Umriß	doppeltrapezförmig
Pfeilform	0°
V-Form	5°
Knickflügel	nein
Wurzeltiefe	1,30 m
mittl. Flächentiefe	1,20 m
Endtiefe	— m
Flügelstreckung	8,33
Trapezverhältnis	—
Bruchlastvielfaches	14,00
Beanspruchungsgruppe	—

Profile:

an der Wurzel	NACA 23012
in halber Spannweite	NACA 23012
am Ende	NACA 23012
Höhenleitwerk	NACA 009
Seitenleitwerk	NACA 009
Einstellwinkel am Rumpf	3°
Schränkungswinkel	4°

Zulassung:

zu Windenschlepp	ja
zu Autoschlepp	ja
zu Flugzeugschlepp	ja
zu Kunstflug	ja

Leistungen:

Gleitzahl	1:19,8
Sinkgeschwindigkeit	0,88 m/s
Geschwindigkeit bei bestem Gleitwinkel	81 km/h
Geschwindigkeit bei bester Sinkgeschw.	64 km/h
zul. Höchstgeschw.	450 km/h
Tragflächenbelastung	14,17 kg/m²
Klafterbelastung	1,70 kg/m²

Abmessungen des zusammengeklappten Flugzeuges:

Länge, gr.	— m
Breite, gr. o. Leitwerk	— m
Breite, gr. m. Leitwerk	— m
Höhe, gr.	— m

Bauart und Baustoffe:

Tragwerk	Holzgerippe mit Stoffbespannung
Rumpfwerk	Schalenbauweise in Holz
Höhenleitwerk	Holzgerippe mit Stoffbespannung
Seitenleitwerk	Holzgerippe mit Stoffbespannung
Fahrwerk	festes Einradfahrwerk

FLUGTECHNISCHE FACHGRUPPE
an der Technischen Hochschule Stuttgart

Technical data of *Felix Kracht's FS 17*.

petitions. On the other hand, I do not know of the exact operation plans for the *DFS 228*, but it appears that the *RLM* would have used it in some ultra high altitude aerial photography.

MYHRA - Did *DFS* ever test *Kracht's* cockpit to see if it would have worked as he designed it? I understand that the first prototype of the *DFS 228* was not made flight ready until 1943, and then it was given a civilian registration number of *D-IBFQ*. I have been told that two additional *DFS 228s* were completed by May 1945. Is this correct, and if so, who built these two *228s*? I have heard, too, that it was *Werde*?

ZITTER - No, *Kracht's* cockpit, as you call it, was not tested. We had no way of getting one of the *228s* up to an extreme altitude and then releasing the cockpit from the center fuselage. *DFS* built all three prototypes, but *Werde* was to build the *228* in series at Freliessing, a village a little north of Ainring which is near Salzburg.

MYHRA - Well, *Kracht's* detachable cockpit worked. It was tested in the Soviet Union in the early 1950s. In fact, it saved a pilot's life after he experienced uncorrectable problems with the *DFS 346*. That pilot was *Wolfgang Zeise*, a former test pilot for *Siebel Flugzeugbau* and later a reconnaissance pilot flying *Arado Ar 234B-2s* out of Oranienburg, northwest of Berlin. On the other hand, I have heard that the *DFS 228's* retractable landing skid was built by *Schmetz* of Herzogenrath, and also the machine's control surfaces.

ZITTER - Yes, they built the landing skid for sure. I am not sure anymore about the surface controls, however.

MYHRA - Would you know the aspect ratio on the *DFS 228?* I have it as 9.5 to 1. Also, do you know who might have built its wooden spar?

ZITTER - The aspect ratio was somewhere near 10 to 1, I recall. As far as the wing's spar, it was a laminated wooden spar, and I do not know for sure who built it, whether *DFS* or some other *werk*. At that time of the war there were many, many little shops building things out of wood for the *Luftwaffe* and the aviation industry.

MYHRA - *Felix Kracht* told me that at this time he was also in charge of some small furniture shops who were making wings for the *He 162* "peoples' fighter," and that this small wood shop was near Ainring.

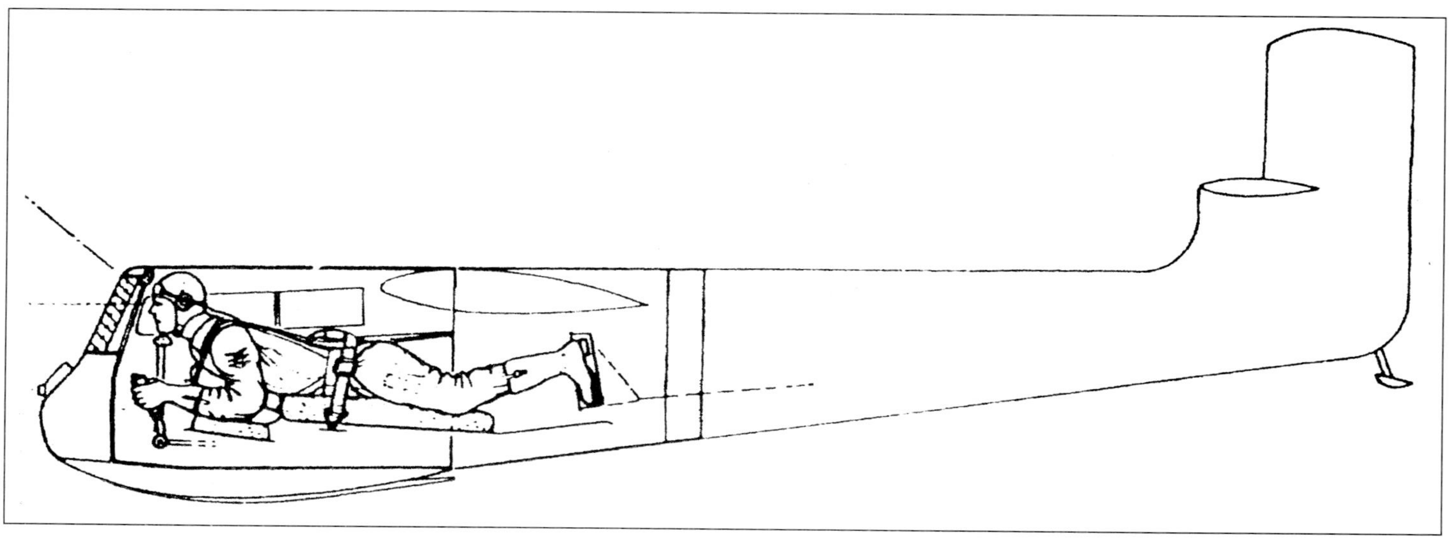

Another example of a prone piloting position aircraft was the twin engine trainer known as the *Berlin B-9*. *Felix Kracht* planned to utilize the *B-9* as a training familiarization aircraft for pilots going on to pilot the *DFS 228* and *DFS 346*.

The pilot of a *B-9* is shown here being assisted prior to take off.

ZITTER - That is correct, but the shop was really closer to Nueburg, near Wien, Vienna.

MYHRA - I have heard, too, that in the summer of 1945, the *DFS 228 V1* coded *IBFQ* was given to the British by the Americans, and then they transported it to England to study.

ZITTER - It is possible, but I do not recall the circumstances concerning its removal. Actually, the Americans found both surviving *DFS 228* machines. They checked out their structural design, method of operation, performance of the propulsion unit, and fuel systems. As I recall, what was of particular interest to the Americans was our jettison cockpit cabins on the *228 V2* and *V3*. That was about the only real difference between the *228 V1* and *228 V2/V3*. So it appears that the Americans kept the most important specimen, the *V3*'s prone position cockpit cabin, and gave the *V1* with its conventional pilot seating position to the British.

MYHRA - Yes, this sort of horse trading happened frequently. Although America and England were Allies, American self-interest ruled. Then, after all, the Americans pretty much won the war. Did you happen to be present at Ainring when the Americans came and removed the *DFS 228 V1* to Stuttgart for safe keeping?

ZITTER - For the last year of the war (1944-1945) I was stationed at *DFS*-Ainring. The *DFS 228 V2* was not in Ainring, but had been in Hörschuning...and I think that it was lost due to war damage on a crash landing. But at that time (1944-1945) we had the *DFS 228 V3* in Ainring. So it had to be our *228 V3* which the Americans took away. I know this because many things on the *V3* were changed as

A pilot is seen in the process of entering a *B-9*. Notice that the pilot wears his parachute about level with his stomach. Actually, the pilot used the parachute pack as his couch while piloting the *B-9*.

The cockpit of a *B-9*. Notice at the bottom of the photograph what appears to be a container...this was the "tub" where the pilot's parachute pack was placed during flight and which served as his cushioned couch.

a result of its flight testing, and for this reason it was still at Ainring in May 1945.

MYHRA - Yes, I have heard, too, that the *DFS 228 V2* suffered war damage and was lost. I have been told that the 228 *V3's* unfinished cockpit cabin was taken by the U.S. military intelligence people, too. It appears that the *V3* was not brought to the USA, but got only to Stuttgart, and after inspection, it was destroyed. Then, again, maybe only its jettisonable cockpit cabin was sent to America and the rest of the fuselage and wings were scrapped.

ZITTER - I don't know, either. You should talk to a man by the name of *Zucker* who was an assistant to *Professor Ruden*. At *DFS Professor Ruden*, who is now dead by the way, was the leader of *DFS'* Aerodynamic Branch and Project Bureau. *Zucker* might know of the *228 V3's* details.

MYHRA - The *DFS 228 V1* had the pilot seated in a conventional manner, but this was later changed to the prone seating position? When?

ZITTER - In the *228 V2* and *V3*.

MYHRA - Did you know that the DFS 228 *V1* finally ended up at the Royal Aircraft Establishment (RAE), Farnborough, England?

ZITTER - That, too, is possible, but I am not aware of any of the details.

MYHRA - *Felix Kracht* told me that immediately after the war he met and traveled for a time with *Col Charles Lindbergh*.

ZITTER - Yes, that is so. *Lindbergh* came to Ainring about two or three days after our 8 May 1945 surrender. It was for sure some time between 8 May and 1 June 1945. This was the time that the American forces were occupying the Ainring area. In fact, I was with them as the driver of the American *Jeep* we all traveled in.

MYHRA - Why were no powered flights made in the *228?* Were the *HWK 509D* rocket drives intended for the *228* ever tested on the ground?

In the USSR postwar, the captive German aviation specialists built their own *FS 17* in order to train *Zeise*, *Trauter*, and others the techniques involved in prone position piloting the *DFS 346*. A German pilot is shown inside the *FS 17*.

The cockpit of the *FS 17* sailplane constructed in the USSR postwar.

ZITTER - Yes, tested on the ground a lot, but I don't know why there were no powered flights. We never flew a single *228* under its own power, however, perhaps as many as 40 or more gliding flights were made.

You have asked about the people behind the *228* and *346*. I do not know who sponsored these research aircraft. *Oberst Siegfried Knemeyer* I had met before. My first contact with *Knemeyer* was in *Erprobungsstelle*-Rechlin in 1941/1942. In October 1946, we both were on the same American military airplane for a flight to Paris. He quit, and in mid 1947 was hired by the American government, flew to America, and I have never seen or talked to him again.

MYHRA - *Knemeyer* discharged all the personnel in his *RLM*-Berlin department on 23/24 April 1945. He and *General Baumbach* left Berlin by auto to Hamburg to meet with *Speer*. On 28 April they were requested to appear before *Himmler* at Mecklenberg. *Himmler* told them that he was going to do all the post war negotiations with the Allies. On April 30th the two men were at *Karl Dönitz's* headquarters at Eutin, which was about forty miles from Hamburg, when the Soviets announced that the *Fürher* was dead. Several days later he and fellow officers with *Dönitz* were captured by the British moving on Hamburg. Upon his capture, he was transferred to an English prison camp at Münster, and then on to a camp at Latimer, England. *Knemeyer* was transferred to an English prison camp in Belgium and was finally released on 21 May 1946. He accepted a position offered to him by the *Arsenal de Aéronautique*, Kressbroun, Paris. Then, in July 1947, *Knemeyer* resigned from the *Arsenal* and became an employee of the U.S. Government at Wright-Patterson Air Force Base, Dayton, Ohio.

End of *Hermann Zitter* Interview

Hans JACOBS
15 May 1987
Siegen, West Germany

MYHRA - I have recently finished a biography of the *Horten* brothers, and now I am working on a book on the history of the *Mach 1.0* research aircraft and ultra high altitude photo reconnaissance machine known as the *DFS 346*.

A port side view of a pilot in the *FS 17* sailplane built and flown in the USSR postwar. It appears that the overall roominess inside the cockpit was much greater than the initial *FS 17* built in prewar Germany.

The Soviet postwar version of the *FS 17* appears larger than the prewar German sailplanes. Notice the wing span on this prone position sailplane built in the USSR.

JACOBS - Well, *Lippisch's Me 163* rocket plane was perhaps the first aircraft to reach *Mach 1.0* in a dive.

MYHRA - I would like to obtain a little more history of the *DFS* from you today, and also some information about *Professor Georgii*.

JACOBS - I can only tell you in general about *Georgii*.

MYHRA - Here is a photograph of the so-called supersonic research aircraft, the *DFS 346*, under design at *DFS* and later under construction at *Siebel*-Halle. It was intended that it would reach *Mach 1.0* in a climb.

JACOBS - I did not participate in the *DFS 346* project. In fact, this aircraft is the flying machine from *Felix Kracht* at *DFS*-Ainring. Getting back to *Georgii*, in 1925, he joined an organization called the *Rhön-Rositten Gesellschaft* (*RRG*). From this *RRG* was the origin of the *DFS* in 1939. *Professor Georgii* then became the *DFS'* chief. The relation between *Georgii* and the *Reich's-Verkehrs-Ministerium* (*RVM*) was especially good, and this relationship extended back to *WWI*.

MYHRA - Did Georgii participate in WWI?

JACOBS - Yes, he was a meteorologist in the aviation corps. That was where he met *Brandenburg* and *Baeumker* and became lifelong friends to both. *Baeumker* later became director of research and development within the *RLM*.

MYHRA - Was *Baeumker* a fighter pilot in *WWI*?

Although a prone position flying machine, the *DFS 228* was to have a strict business use about it...ultra high level photographic reconnaissance. Shown is *Theodor Rowehl*, the highly decorated leader of the *Luftwaffe* photo reconnaissance group. *Rowehl* was insisting on a new dedicated flying machine able to reach ultra high level reconnaissance. The *RLM* offered him the *DFS 228* with its twin infra-red camera's.

Another highly decorated aerial photo reconnaissance pilot was *Siegfried Knemeyer*. Later in the war, *Knemeyer* would take over *Udet's* old job at the *RLM* from *Milch*. When this happen in 1943, efforts to develop an ultra high altitude photo reconnaissance machine was being given serious consideration. *Knemeyer* is shown standing in the cockpit of the all-wing *Horten Ho 3d*.

JACOBS - No, not a fighter pilot, but also as a meteorologist in a *luftballoon* group. Both *Professor Georgii* and I were very close friends of *Hanna Reitsch*.

MYHRA - I believe that *Baeumker*, while he was in charge of Research and Development in the *RLM*, was very instrumental in deciding to get the *DFS 346* designed and built. Many of us feel that he was the "man" behind the *346*. How did *Baeumker* come to support *Mach 1.0* research so late in the war?

JACOBS - Well, it is my opinion that it had nothing to do with the *RLM's* own current interests, but was more like something colleagues of *Baeumker* in the *RLM* wanted to do...war, or no war. Maybe they wanted bragging rights post war that the Germans were the first to break the speed of sound in level flight...before the Americans, British, or Russians. That is what I think. You should not try to make too much more out of it...to make it more complicated than it really was. The *DFS 346* appears to me to be an ego fulfillment of a bunch of powerful men at the highest ranks within the *RLM*...to build themselves a play toy and fly it before they all had to surrender to the Allies. Knowing these people as I do, this is really the reason...the only reason. *Baeumker* probably just went along with the guys and their play ideas. You see, *Baeumker* just sort of naturally supported *Georgii's* on-going interests between 1933 and 1941, when it was my friend *Alexander Lippisch* who, at the *DFS*, was highly interested in rocket powered aircraft, including the *DFS 346* and its predecessor, the *DFS 228*. That's about all I can suggest regarding the *346* project.

MYHRA - Tell me a bit about the man *Georgii*, his work habits, relationship with others, and so on. He seemed to have interests in all forms of aviation research.

JACOBS - I believe that *Georgii* was a very noble man. In fact, I worked for *Georgii* between 1927 and 1940. I never experienced a time seeing the man ever getting really angry, but was always very calm. He was very good in raising research monies for *DFS* projects. He was a good diplomat, too. He did not know a great deal about technical matters, but he always helped people if they had an idea. He would encourage them and give them the support needed to achieve their ideas. As the leader of the *DFS* institute, it had really great times under *Georgii's* direction. *Georgii* had a close relationship to staff, colleagues, workers, and to *Ernst Udet*, too.

MYHRA - How did this close relationship to *Udet* come about?

JACOBS - *Georgii* was the type of person who easily acquired friends. He was a person who easily made friends with people. He had a winning way of treating people.

Up, until *Knemeyer's* promotion as head of the *RLM's* Technical Department, the *Luftwaffe* had been using the old *Junkers Ju 86P2* for photo reconnaissance. This twin engine machine with its pressurized cockpit cabin "egg" had regularly flown over the British Isles at 40,000 feet altitude. Next, the *Ar 234B* replaced the *Ju 86P*, however, the *Luftwaffe* needed a machine which could fly even higher than the *Ar 234B*.

A very nice mid-altitude aerial photo reconnaissance machine successor to the *Ju 86P*, was *Arado's Ar 234B*. It could fly higher and faster than the English "*Spitfires*," but it didn't have a pressurized cockpit needed for ultra high altitude photo reconnaissance.

MYHRA - What about *Georgii's* relationship with *Baeumker*? Did it continue on through the years after *WWI* into the 1930s and 1940s?

JACOBS - Yes. The personal relations continued on, but the business relationship ended in 1933.

MYHRA - What do you mean?

JACOBS - I do not know. I never did have any personal relationship with *Baeumker*. I never got to know him. People like *Baeumker* were the "big shots" in Berlin. *Udet* had a very personal interest in sailplanes. *Udet* flew sailplanes himself, and consequently he was always very interested in new types of sailplane planform, construction, sailplane pilots, and constructors. *Udet* got to know the sailplane builders/flyers because he often flew their aircraft. This is how I personally got to know *Udet* so well.

MYHRA - How well did *Georgii* get along with the other aviation research establishments, such as *AVA*, *DVL*, *LEA*, and so on in Germany in the 1930s and 1940s?

JACOBS - He had good, friendly, relationships with these organizations. In fact, these separate research groups would help each other out with their problems a great deal. For example, *DFS* did not have a wind tunnel, but *AVA* did. Then again, the relationship to *DVL* was not all that good because these people felt that their work was much more superior to that of *DFS*. *DVL* said that they worked with "real" aircraft, and not merely with "gliders." There was no

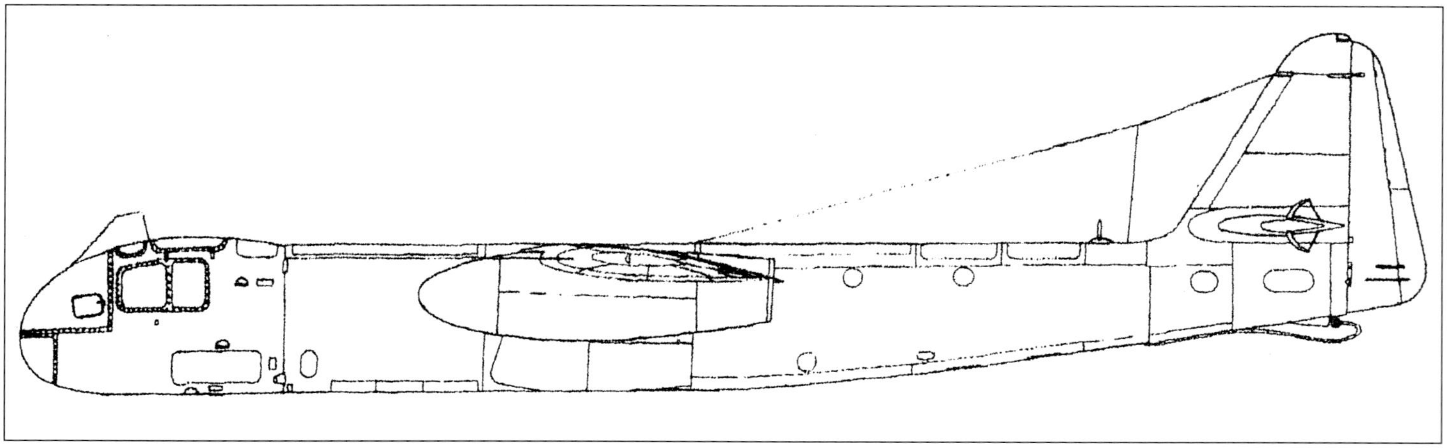

***Arado's* engineers had proposed equipping one of their *Ar 234Bs* with twin *HWK 509B-1* bi-fuel liquid rocket engines to reach ultra high altitudes for photo reconnaissance. Its two *BMW 003As* would have been removed and within the empty nacelles, a single *HWK 509B-1* rocket engine installed in each.**

The *HWK 509B-1* bi-fuel liquid rocket engine appearing as it would look installed in the nacelle occupied by the former *BMW 003A* turbojet engines on the *Ar 234B*.

personal animosity, but it was just that *DVL*, as a whole, looked down upon *DFS*.

MYHRA - In the 1930s, who was the main supporter of *DFS*? Was it *Ernst Udet*?

JACOBS - Yes, I would say that it was mainly *Udet*, and since he was friendly toward *DFS*, then also the rest of the *RLM* in general was sportive and friendly, too. After *Udet's* death, one of *DFS'* research institute's names was changed to include *Udet's* name. *Professor Georgii* was very renown internationally, too. We can consider him one of the fathers of German modern day high altitude gliding. *Georgii* made efforts to see that the glider people obtained an international facility for gliding at the *Rhön/Wasserkuppe*. Thus, all nationalities participated. *Georgii* also wanted to see that gliding became an Olympic event. I would have participated in the 1940 Olympics in Tokyo, but the games were never played. *Georgii* saw to it that people who would participate in upcoming 1940 glider activities at the Olympic games had to build their own gliders for this event. One of my gliders, which was called the *"Mieser,"* had been selected, and we looked forward to going to Tokyo in 1940. I think now that examples of my *"Mieser"* have been built over 100 times in postwar France and England.

MYHRA - Did the *DFS* divide in 1933 into two areas of interest: one for gliders, and the other for high-speed aircraft?

JACOBS - Yes, after 1933, *DFS* more or less split into two parts. When *DFS* moved to Darmstadt in 1933, this was the branch for gliders, and it was at this time I left *DFS*. Already in 1927, I had designed and built the first powered airplane without a tail. Then came the *Rhön/Wasserkuppe*, and four different aircraft were tried out, such as *Lippisch's* "*Delta I*" and "*Storch*," one by *Gunter Preis*, and an aircraft by *Herman Am*.

MYHRA - It appears that *Georgii* felt comfortable with all sorts of aircraft and with different planforms. He seemed always ready to try something new, something on the cutting edge of technology, especially the cutting edge of technology. He seems to have been a very liberal person and not at all stuck in conservatism. All types of aircraft, from exotic to old standby's were of interest to him.

JACOBS - Back at the *DFS*, there had been a dispute between *Lippisch* and *Georgii* because *Georgii* had accepted the construction of my *DFS 230*. *Lippisch* was bitter about it, and it was one of the reasons why *Lippisch* went off to *Messerschmitt AG*. *Lippisch* was a person who always had to have things his way. If something was done which was not according to his wishes, well, he became immediately angry. *DFS* really did not have the capacity to build high speed aircraft the likes of which *Lippisch* wanted to get into. So, when *Lippisch* moved to *Messerschmitt AG*, well, it was thought to be kind of a good choice, because the *Messerschmitt AG* was more able to handle the construction of high speed aircraft.

MYHRA - If *DFS* was doing all this research into high-speed flight, why were funds not forthcoming for research and prototype con-

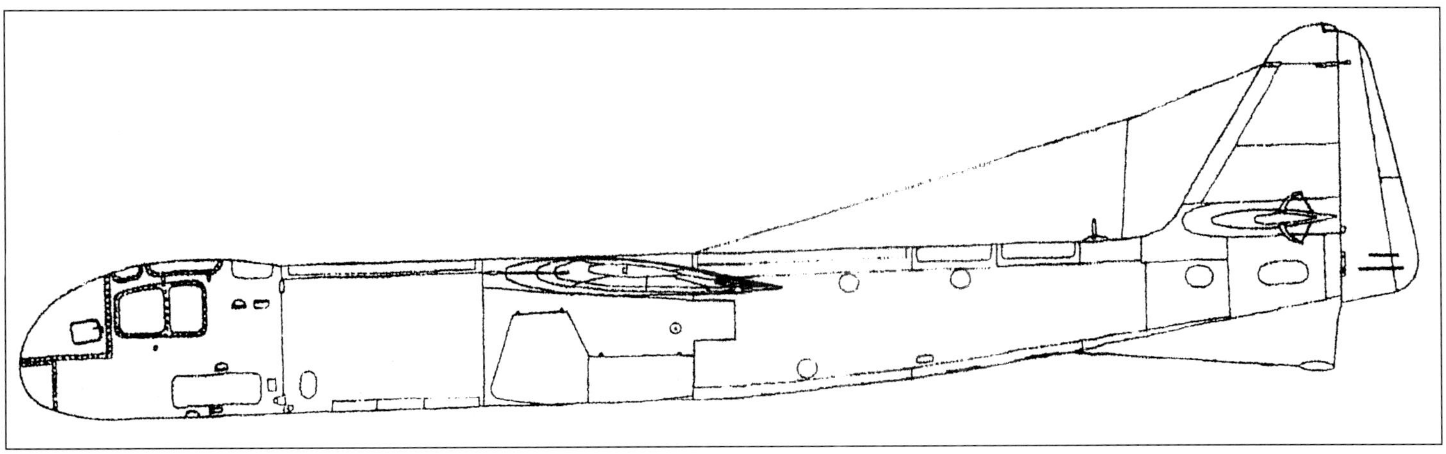

A second ultra high altitude proposal by *Arado* engineers was to equip a *Ar 234B* with two *HWK 509C* bi-fuel liquid rocket engines in place of its *BMW 003A* turbojet engines. The *HWK 509C* had two combustion chambers: a top or upper main chamber and a lower cruising chamber.

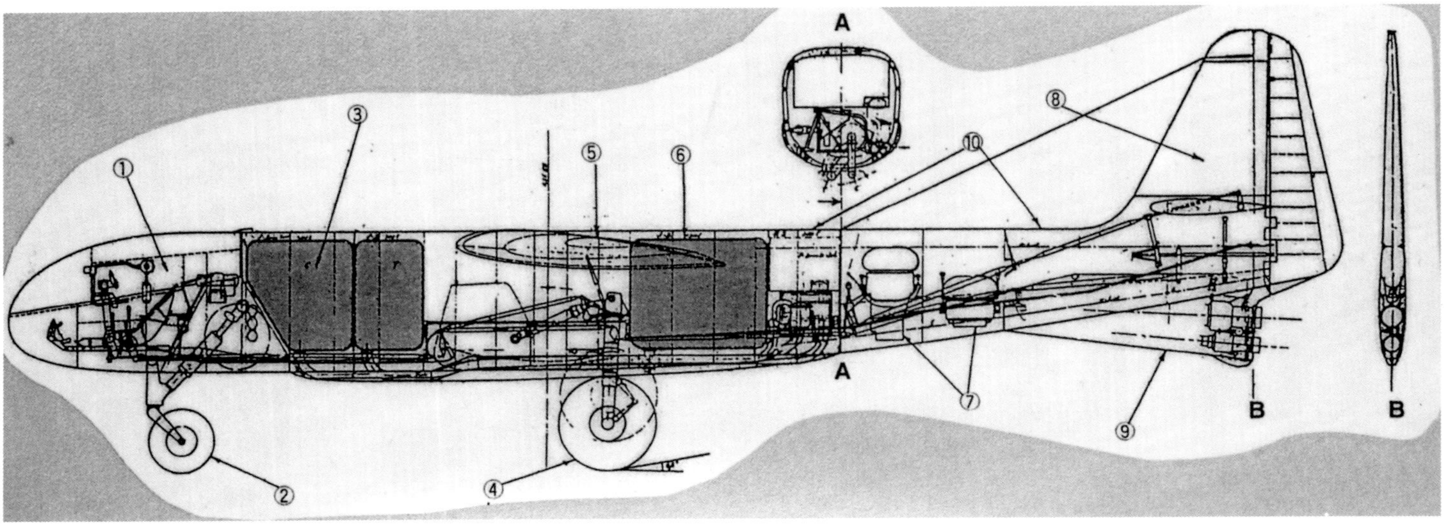

A pen and ink drawing from *Arado* featuring internal arrangement from a port side view of their proposed ultra high altitude photo reconnaissance replacement for the *BMW 003A* powered *Ar 234B*.

struction, as well? Why was it necessary to move on to *Willy Messerschmitt*, a man much like *Lippisch*, knowing that these two personalities would eventually come in conflict with one another?

JACOBS - The support for the *Me 163A* and *163B* was good. However, there had been an internal dispute with the *RLM* over aircraft with or without a tail. What was the better choice? The big and famous constructors all thought that a tailless was not good enough for maneuvering in fighting situations.

MYHRA - Yes, the *Horten* brothers suffered the same criticism, and pretty much the same fate.

JACOBS - Because most big constructors did not think very highly of tailless aircraft, the *RLM* didn't want to fund *Lippisch's* research and development anymore, either. *Lippisch*, when he was still with *Messerschmitt AG*, tried to change the attitude of the *RLM* toward his tailless aircraft by asking *Ernst Udet*. He did not do it directly, but indirectly. *Lippisch's* efforts to change their funding priorities on tailless aircraft made *Willy Messerschmitt* really mad. On the other hand, *Lippisch* claimed that he was not able to go forward with *Messerschmitt*. Well, because of this quarrel between the two titans of the aviation industry, there was a big conference at the *RLM* in 1943, over what should be the future of tailless aircraft. *Wolfgang Spate* wrote about it in his book, especially on page 143.

MYHRA - Yes, at this conference the *Horten* brothers' all-wing ideas were criticized, too, and so both *Lippisch* and the *Hortens* were ganged up on. The whole flavor of the conference was really to put down tailless and all-wing aircraft designs. They did it, too.

JACOBS - Yes. *Lippisch* made two major mistakes in his career. First, he thought of constructing only all-wing aircraft, with passengers as well as its engines located inside the wing. It did not work. His second big mistake was that *Lippisch* thought mainly of the tailless configuration, and this planform didn't work either. His only idea which was right and which has been put into widespread practice has been his use of the swept back wing.

MYHRA - What did *Professor Georgii* think of all this infighting among these two strong German aviation personalities? What planform did he favor, tailless or conventional? Where did he stand in this controversy?

JACOBS - *Professor Georgii* didn't really understand the technical arguments either for or against. As a result, *Georgii* tried to

The twin combustion chambers (over and under) of the *HWK 509C* bi-fuel liquid rocket engine proposed for use in the *Ar 234B*. It was never built.

remain neutral, because he didn't understand any of the pros or cons of it all. He was not against *Lippisch* because he was a tailless advocate. *Lippisch* was a very difficult man to handle, anyway, even in the best of times.

MYHRA - I think that it would have been very predictable that *Lippisch* and *Messerschmitt* would not have worked well together. Both were very difficult men. Both pretty much prima-donnas, both highly intellectual, pioneers in their own right, and to make matters worse, both strong willed.

JACOBS - Yes, difficult for him because *Willy Messerschmitt* at that time did not believe in the suitability of a tailless machine.

MYHRA - So, *Georgii* was able to disassociate himself from this clash of the two titans? But how was he able to still walk the middle line without angering or offending both sides?

JACOBS - Yes. *Georgii* was a man who tried to evade quarrels and disputes. I think that *Lippisch* quarreled with *Georgii* over the fact that he had not raised enough funds for tailless research projects. But these were internal matters at *DFS*, and I do not know the full details.

MYHRA - So then, *Georgii* would not tend to take a stand one way or the other?

JACOBS - It was very hard for *Lippisch* to understand and accept *Georgii's* lack of support or taking a position on his tailless planforms.

MYHRA - After *Udet's* suicide, who then became the chief sponsor or supporter of tailless inside the *RLM*? Who continued to support *DFS*?

JACOBS - It was *Erhard Milch*.

MYHRA - How was *DFS* then looked upon by the *RLM*: as a research arm of the *RLM*, or a constructor of special projects?

JACOBS - Well, there were two other people in *RLM* largely responsible for support to *DFS*, and these men were *Lucht* and *Reidenbach*. Regardless of the death of *Udet*, the *DFS* was still supported. You see, the *DFS* was not just a branch of the *RLM*, but it continued to be an independent research organization.

MYHRA - Did the *DFS* obtain all its funding from the *RLM*?

JACOBS - Yes, 100 percent. Another big project within the *DFS* was the "piggyback," one aircraft carried by another. *Felix Kracht* did a lot of work on this concept. A second big project was an experiment to stop airplanes while landing through the use of parachutes.

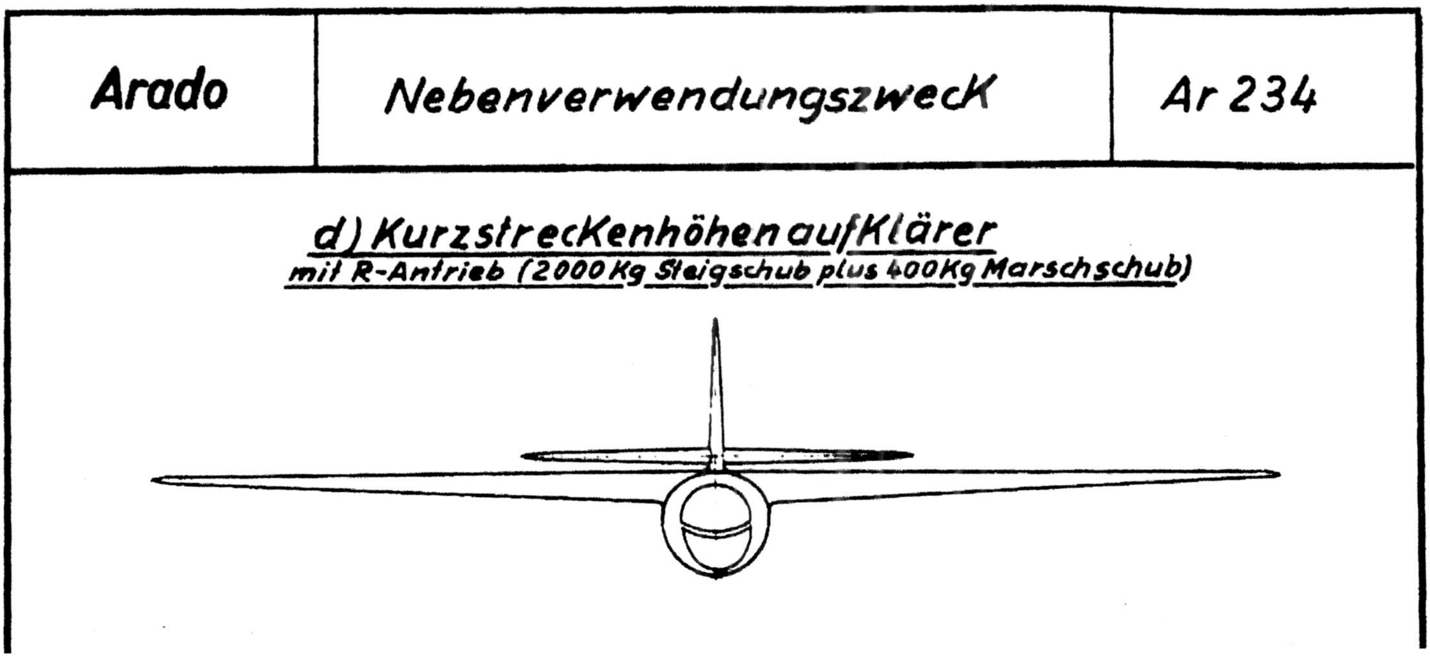

A pen and ink drawing from *Arado* featuring a nose on view of the slim profile of their proposed *HWK 509B* or *509C* powered *Ar 234B* ultra high altitude photo reconnaissance aircraft with its low aspect ratio wings.

D-I
BFQ

A direct tail on view of the *DFS 228* and the thrust nozzle for its single *HWK 509D* bi-fuel liquid rocket engine. Notice, too, that its hydraulic articulated metal landing skid is fully extended. Digital image by *Mario Merino*.

MYHRA - Did you know *Oberst Knemeyer*, of the *RLM* Technical Office? After *Udet's* death, *Milch* took over most of his duties. But in about 1943, *Siegfried Knemeyer* was assigned the task of looking after R&D in the *RLM*. It was at this time that *DFS* started getting into more exotic aircraft research.

JACOBS - I did not know this man *Knemeyer*. I may have met him, but I don't even know the name.

MYHRA - Are you familiar with the *DFS 228*, the single engine *HWK 509D* bi-fuel rocket-powered research reconnaissance sailplane by *Felix Kracht*?

JACOBS - Yes, I am aware of this aircraft, but I didn't have anything to do with its design or prototype construction.

MYHRA - Did any of the ideas for the reconnaissance function of the *DFS 228* came from *Eugen Sänger*.

JACOBS - Yes, I knew *Sänger*, but the designer and constructor of the *DFS 228* was *Felix Kracht*.

MYHRA - Well, *Sänger* had this idea for a high speed, high altitude reconnaissance aircraft. It involved a powering up and gliding down, then powering up again. Sort of a sawtooth approach. *Knemeyer* supported the idea, too, and he had achieved fame for his daring reconnaissance flights over the USSR and received a *Knights Cross* from *Göring* for it. Anyway, it was getting to the point where a great many exotic research jobs were being given to *DFS* through *Knemeyer*, chief of the *RLM's* Technical Office, and to *Eugen Sänger*, by *Hans Antz*.

Opposite: The *HWK 509D* powered *DFS 228*. Notice the single opening in the tail for the *HWK's* 3,750 pounds [1,700 kilograms] of thrust. Beneath the combustion chamber can be seen the *228's* metal tail skid. Digital image by *Mario Merino*.

JACOBS - *Hans Antz* worked with *Reidenbach*.

MYHRA - Yes, *Hans Antz* was a member of that same small, powerful group of officers in the *RLM*, including *Professor Seewald*.

JACOBS - Yes. *Seewald* was the chief of the *Reich's* Secretary of Traffic Department.

MYHRA - So, there was this group of people coming together in 1943 and 1944, who were interested in exotic research, exotic power plants like the pulse-jet, ram-jet, bi-liquid rocket power, and so on, to help them obtain ultra high speed and ultra high altitudes. So, I'm wondering where this idea for a *Mach 1.0* research aircraft fit into this collection of men at *RLM* and *DFS*?

JACOBS - Maybe. But initially the *DFS* worked on "peace" type projects, sailplane research, and the military thought that *DFS'* projects would not be effective during wartime or really effect the outcome of the war.

MYHRA - Yes, I agree that many of *DFS'* projects by 1944 and 1945 were merely carried on and out through the whim of powerful supporting individuals within the *RLM*, and frequently without any thought to their war effectiveness. It's as if someone on a whim said "let's try this and see if it works." In fact, some of these exotic pieces of research may have only accomplished the taking away of war materials which might have been better used elsewhere.

JACOBS - Yes. A lot of funds were taken away for these projects which could have been better put to use elsewhere. Too many projects were started, and the *Me 163* was one of those projects. It had nothing to do with the outcome of the war. It accomplished absolutely nothing.

MYHRA - Did you know a man by the name of *Helmut Schelp* in the *RLM*? He was in charge of all turbojet research, planning, and development between late 1938 and May 1945?

JACOBS - No.

MYHRA - Well, he lives now in California. But in Germany during those years he was a very powerful person in the *RLM*. I asked him how was it that it came to be during the war that the *DFS 346 Mach 1.0* research aircraft came to be? How was it that this project actually saw the light of day during those desperate times of late 1944 and into early 1945? *Schelp's* answer was something like this: "There was about one dozen of us men in the *RLM*, and we ran the show! We could do what we wanted to do. Some of these projects were carried through because so-and-so merely wanted to do it. Since we ran the show we did what we wanted. The members in this "club" had their whims and personal wishes carried out. If someone wished to have a machine that would reach *Mach 1.0* and beyond, well—which *Hans Antz* wanted to do by the way—the others in the group said... sure, why not...go and have it built and do it." *Hans Antz*, by the way, was heavily involved in the development of the *Me 262*.

JACOBS - Before the war, industry and research offered things to the *RLM*, for example, the *DFS 230*. But in the war it was different. The *RLM* asked them to do this and that, but also industry told them that we were able to do this and that, so give us money.

MYHRA - Technology was available so that they could do certain things, like high speed research, and I'm sure that the people in industry said "wouldn't it be neat to go super fast," and the people

The *DFS 228* as seen from its tail port side. It appears gliding back home since no visible thrust is coming from its single *HWK 509D* nozzle. Digital imagine by *Mario Merino*.

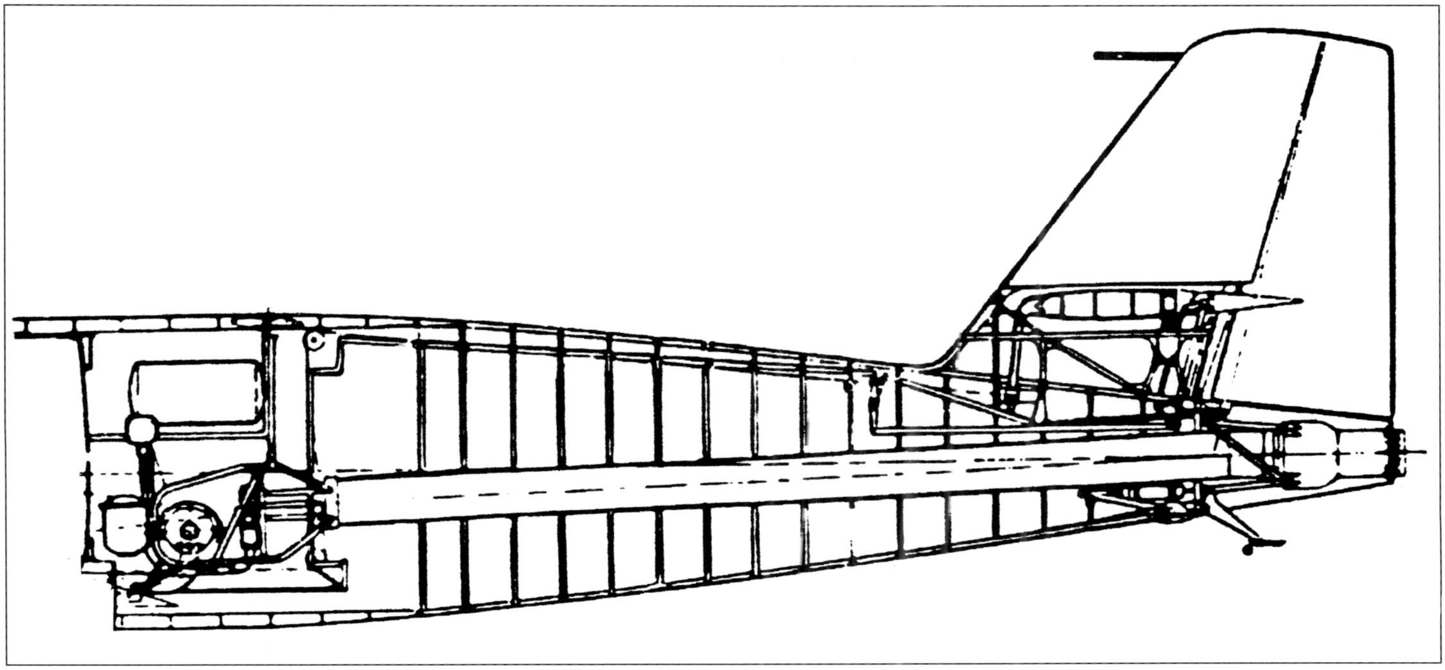

A pen and ink drawing from *DFS* featuring the mid fuselage location for its single *HWK 509D's* main engine (left), thrust tube, and combustion chamber (right) in the *DFS 228.*

in the *RLM* said "yes, that would be a neat idea." Whether it had military value or not was not the issue. It would merely be nice to try it. So, industry collaborated with people in the *RLM*. In general, there was this little club of people from industry, from the research labs, the universities, and then from the *RLM*. It appears that they then all used the *DFS* for their vehicle to realize their personal dreams.

JACOBS - No, not entirely. *Georgii* had his own ideas, too. He'd say to me that we were going to *RLM* because we needed funds. It was not that the *DFS* was told to do everything. Each researcher in the *DFS* had his own ideas, too and they then would ask the *RLM* for money. Most people at this time, whether or not they were employed at the *RLM*, in the aviation industry, or in research, were all too young and too inexperienced, and they did not have the real mission and knowledge for the research they did. Take *Ernst Udet*,

The typical *HWK 509B-1* bi-fuel liquid rocket engine. It was a modular yet when bolted together, a one piece unit beginning with the fuel pump and fuel pressurization head, thrust tube, and combustion chamber.

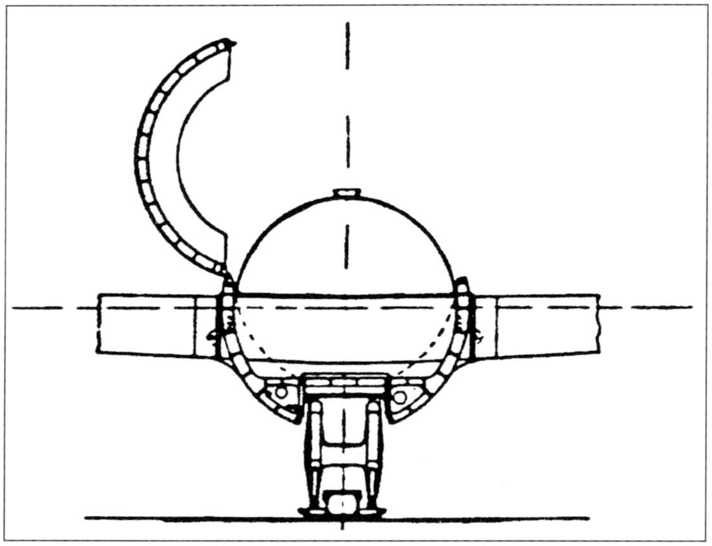

The mid fuselage location for the *HWK 509D* bi-fuel liquid rocket engine head, *C-Stoff* and *T-Stoff* tanks in the *DFS 228*, were enclosed by a one-piece cover which could be opened for access via a hinge or removed entirely.

for example; he was a great flier and a good man, but also absolutely incapable of handing complicated R&D. Even though *Udet* was a very good flyer, this was not good enough, because he didn't know how to manage. The *DFS* and the *DVL* all had solid bases, for they had worked on research projects for years. But the *RLM* was put together very quickly, and all these people were newcomers. Before, most of these men were nothing, and now at age 33 they were "big shots." The leadership was bad. Everything was different in the USA, France, and England, because these research and military organizations had a more solid base with better leadership than our *RLM*.

MYHRA - Yes, the *RLM* was a tragedy, like *Udet being* the wrong man in the wrong job. It drove *Udet* to his death, because he was blamed for many of the aircraft failures of the time, such as the *He 177*, the *Me 210*, the *Fw 190A*, and so on. *Udet* was even being blamed for the failure of the *Luftwaffe* in England.

JACOBS - Yes. I will tell you a little story. *Hanna Reitsch* was my friend, and as a flyer she flew my new aircraft and she had many, many contacts with the "big-shots" in Berlin. Once *Hanna Reitsch* was invited to *Göring's* house for dinner and they talked about lot of things I have spoken about with you today. *Reitsch*, prior to dinner, asked *Göring* about the bombing attacks of 2 February 1944 on Regensburg/Schweinfurt, with its 300 to 400 bombers, and that there had been very little defense by the *Luftwaffe*. The *8th & 15th* USAAF bombers merely came in and dropped their load and left. Plus, it was in daylight, too. At this dinner *Reitsch* asked *Göring* what about all these American bombers coming in during daylight and with no defense for the city? "How shall all this end?" *Göring* answered that he would just have to push a button, and there would be many fighter aircraft which would be available to defend the cities. Thus, the Americans wouldn't be able to do this bombing anymore. *Reitsch* asked *Göring* "what kind of airplanes are these? They must be very miraculous airplanes." Yes, said *Göring*, it is the *Me 163* from *Lippisch*. *Reitsch* responded, "but *Field Marschall*, there are only three: I have flown one of these machines, another is being flown by *Heni Ditmarr*, and the third one is broken, and all there is to be is a series of twenty *Me 163s* now under construction at *Messerschmitt AG*." So *Göring*, not wanting to hear anymore about it, started shouting at his guests, then took his daughter out in the garden for a walk and to play. Whereupon *Frau Emmy Göring* leaned over and said to *Reitsch* "That's always how he behaves when he is told the truth." From this conversation this is why I came early to the conclusion that the *RLM* had no good leadership. Sorry."

MYHRA - The failing leadership of the *RLM* was a tragedy for all the people who suffered and died in this war. Tell me, what kind of abilities did *Felix Kracht* of *DFS* have?

JACOBS - *Kracht* was a very good and able aircraft constructor. He recently retired as a director of the French Air-Bus program. He is a very good man, and is a better designer than myself.

MYHRA - He came to be a *Flugzeugmeister*, which was quite an honor to have.

Due the overall diameter of the *HWK 509D's* pump head, it had to be placed mid fuselage in the *DFS 228*. As a result, the combustion nozzle had to be separated from its thrust tube and placed at the terminus of its fuselage tail as shown in this photo of the *DFS 228 V1's* combustion nozzle.

JACOBS - I do not know anything about that. But *Reidenbach* was one of the leading men in the construction department of the *RLM*. There were about four different departments: one for air frames, one for power plants, and so on.

MYHRA - How did *Kracht* handle all these research tasks he was given? What of *Georgii*? He must have felt in time that some of these tasks were just plain nonsense. What did he feel about the lack of leadership at the *RLM*?

JACOBS - *Kracht* never uttered personal opinions, because during wartime it was just too dangerous. Technically, *Georgii* did not know too much about such things, whether these requests were good or nonsense. It is very hard for people who did not live at that time to imagine what it was like to work under these wartime conditions.

MYHRA - I have talked to *Otto Fuchs*, and he said he was always in trouble with the *Gestapo* because he refused to be silent. He still tended to speak his mind.

JACOBS - *Otto Fuchs* was a very good Democrat. I also had to face the *Gestapo* twice, so I know what risks *Fuchs* took and what he means.

MYHRA - What did you do after you left the *DFS* in 1940?

JACOBS - I started my own firm for glider construction with a partner by the name of *Swire*, and we called our firm the *Jacobs-Swire Flugzeugbau*. We built some gliders. *Swire* had a wood products company. Later, we constructed guide engines and other military aircraft parts out of wood.

MYHRA - Have you written any memoirs?

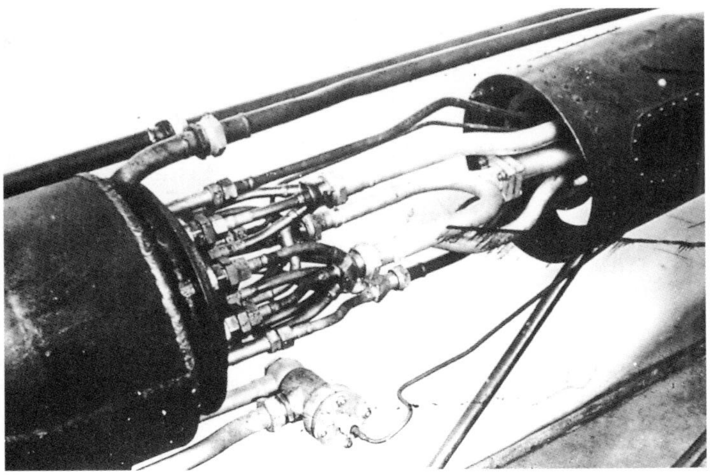

It was risky business to separate the combustion chamber from the *HWK 509D's* thrust tube. Extra long fuel lines had to replace the short ones. Plus, the combustion chamber was separated from its thrust tube and had to have its own support brackets secured to a strong bulkhead.

The single combustion chamber from the *HWK 509D* bi-fuel liquid rocket engine like this was installed in the tail of the *228's* fuselage.

JACOBS - Yes. two volumes, but only for my children to read. I begin by asking the question of how a 26 year old college-educated man could have joined the *Nazi Partei*. I have read every book available on national socialism, and in writing my memoirs I wanted to explain to my children and grandchildren how it was all possible, and how we were so stupid during this period of our lives.

End of *Hans Jacobs* Interview

Hans NIESS
28 May 1987
Bayerisch G'Main bei Bad Reichenhall, West Germany

MYHRA - I have been told that you were one of *Felix Kracht's* assistants at *DFS*-Ainring?

NIESS - Yes, I was. I was in charge of detail construction on the *DFS 228* and on the *Me 328*, with its twin *Argus* pulsejets, but I had nothing to do with the *Mach 2.0 DFS 346* project. *Messerschmitt AG* wanted their *Me 328* prototype design to be built not of aluminum, but out of wood and steel. Yet *Messerschmitt AG* had no experience in constructing aircraft out of wood and, therefore, *DFS*-

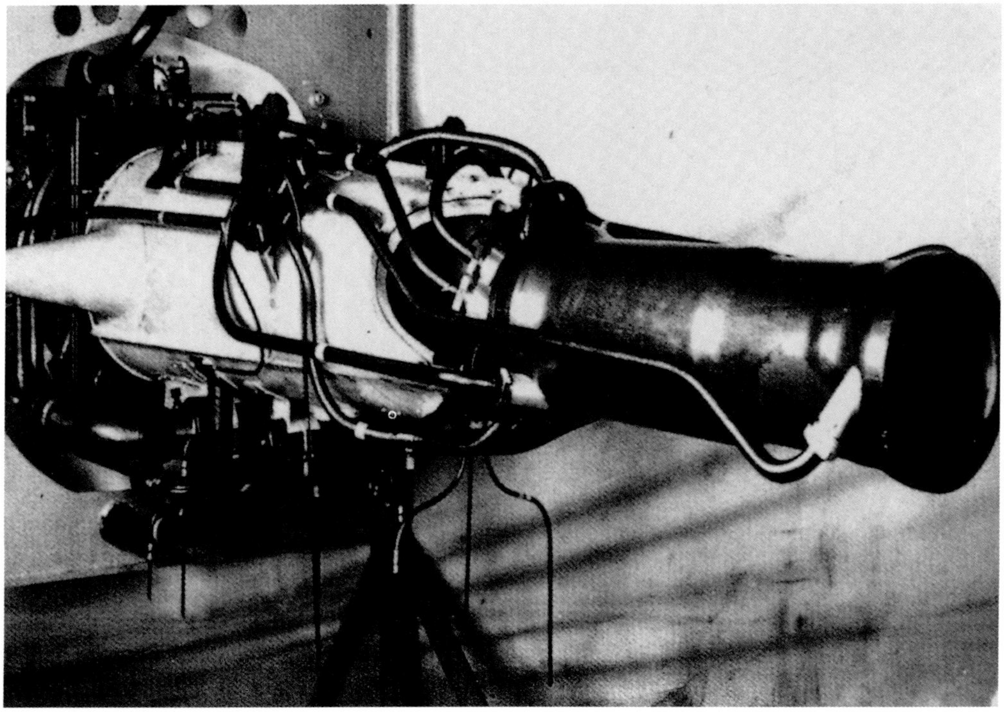

The new bi-fuel liquid rocket engines under development at war's end by *BMW's Helmuth Zborowski* would not have fit in the *DFS 228*, either. These units even had a much shorter thrust tube...in fact the combustion nozzle on this *BMW 3390A* prototype was virtually aft the mixing and pressurization chamber (head).

Ainring had been told by the *RLM* to build the entire *Me 328* prototype out of wood and steel.

Our *DFS 228* was an experimental ultra high altitude reconnaissance aircraft. Its fuselage was mainly wood. The wing was of wooden construction, too, with plywood skin with an aspect ratio of 10. If an emergency should occur at high altitudes the entire cockpit cabin could be separated from the fuselage by the firing of four explosive bolts designed by *Karl Bütter* of He*inkel AG*. The cabin then dropped to a lower, safe altitude by parachute. When this happened, the pilot could eject the cockpit couch out of the cabin. He would then release himself from the cockpit couch and parachute to the ground.

MYHRA - Tell me how you personally got started in aircraft design work?

NIESS - I really got started while I was a student at the Stuttgart University with the so-called *Akaflieg*-Stuttgart. At that time we students constructed many aircraft. I recall one such aircraft was capable of withstanding up to 8 times gravity while flying the aircraft in a prone position. In this position, compression of the blood was not so bad, and it did not run out of the head and down to the feet. This was the first research test flights we made with this machine. We students at the *Akaflieg*-Stuttgart had also designed a prone-flying machine powered by two piston engines, and it was later called the *Berlin B-9*. Our fellow students at the *Akaflieg-Berlin* constructed it for us because there were too few students at the *Akaflieg-Stuttgart*. Most of our students had been taken by the *Wehrmacht*. So, our experimental research design was built at the *Akaflieg-Berlin*. Since it was decided that *Akaflieg-Berlin* was doing the construction, the machine would be known as the *Berlin B-9*.

MYHRA - The Soviets found a *B-9* at *Henschel Flugzeugbau* at war's end, and it was taken to the USSR. Do you know why the *Berlin B-9* was at *Henschel* in the first place?

NIESS - It was because *Henschel* wanted to build a twin engine aircraft with prone pilot seating, and they were studying the *Berlin B-9*'s cockpit arrangements.

MYHRA - Are you thinking of their prototype single turbojet engine powered *BMW 003A, Hs 132?*

Another version of the *BMW 3390A* bi-fuel liquid rocket engine under development at war's end.

The *DFS 228 V1 D-IBFQ* survived postwar. It was found in its hangar at *DFS*-Ainring about the time of Germany's surrender on 8 May 1945, by American Army troops. After its wings were dismounted, the *V1's* entire fuselage with its tail assembly, was placed on this U.S. Army flat bed trailer and removed to the Army's Intelligence collection center near Stuttgart for evaluation.

A poor quality photo of the final known remains of *Felix Kracht* and *Eberhard Meyer's* ultra high altitude photo reconnaissance airplane with the detachable pressurized cockpit cabin *DFS 228 V1*, civil registration *D-IBFQ*. It is not entirely clear where this site was located. It is known that American Military Intelligence gave the *228 V1* to RAE-Farnborough, England postwar. The details of this transaction are not known to this author. The *228 V1* was taken to England by the British for evaluation and study. Sometime in late 1946 or early 1947, the *228 V1* may have been hauled to *Slingsby Sailplanes Ltd* for further examination.

A close-up view of the *DFS 228 V1's* starboard side cockpit cabin. The entry/egress hatch is missing. It appears that the hatch which covers the *HWK 509D's* mixing and pressurized head and fuel tanks, is ajar. It is difficult to see, however, it appears that the plexiglass has been removed or broken out of the several windows.

NIESS - No, it was another *Henschel* prototype machine. However, I do recall their jet powered *Hs 132* aircraft. It was not yet ready for flight at the time of our surrender. The *RLM* intended that the *DFS 228* be flown at extreme altitudes over England on photographic missions. DFS got this job in 1941. Our *DFS 228* would carry no weapons due to its high flight altitude. It was intended that it would be flown between 18 and 23 kilometers [59,000 to 75,460 feet] altitude. For example, the flight plan was that it would be released from its carrier aircraft between 10 and 12 kilometers [32,800 to 39,370 feet] atop the four turbojet engined Ar*ado Ar 234C*. Upon release, the *DFS 228* would immediately power itself up to at least 18 kilometers [59,055 feet], or 23 kilometers [75,460 feet], and there go into a gliding phase, which would carry it for about 120 kilometers [75 miles]. By then it would have dropped down to 12 kilometers [39,370 feet]. At this time its *HWK 509D* bi-fuel liquid rocket motor would be restarted and kept running until the *DFS 228* was back up to 18 kilometers [59,055 feet], or 23 kilometers [75,460 feet], and the process repeated. This power up and glide down pattern did not appear as a "saw-tooth," as some have written, but more like an up-climb and then a long, gradual glide down, an up-climb with a gradual decline, and so on over the course of the *DFS 228's* flight path. The machine was meant to cover about 750 kilometers [466 miles] overall. We at *DFS* were not able to test a prototype in high altitude powered flight, because the cockpit cabin leaked, losing air pressure. We had to build them out of wood, and we could not make the wooden cockpit air tight.

The *DFS 228 V1's* port side wing laying on the gravel. It is hard to tell from this poor quality photo, however, what the wing's condition is in as it lays on the open ground.

MYHRA - Is this the reason why powered flight was never attempted?

NIESS - Yes...the only reason. The *DFS 228's HWK 509D* liquid rocket engine was ready. Its mixing chamber was located in the center fuselage, as were the fuel tanks and *Zeiss* infrared cameras. The *HWK 509D's* combustion chamber and nozzle were located in the fuselage rear. But we had many problems in maintaining cabin pressurization equivalent to 8 kilometers [26,200 feet], and only a few test flights were made, but at low altitude and only as a sailplane. Engineer *Rolf Mödel*, who is shown on the cockpit couch in one of your photographs, had the unpleasant job of building these cockpit couches. On his last prototype cabin, it had a full glass nose. The glass nose, along with the cockpit couch, slid forward to permit the pilot entry and egress. Despite all our work, we could not maintain cockpit pressurization within the *DFS 228's* cabin.

MYHRA - Where was the cockpit cabin constructed, and did *Mödel* build the several cockpits, too?

NIESS - The cockpits were designed by *Felix Kracht* at *DFS*-Ainring, and *Mödel* constructed them all out of wood. I believe that several of the later versions from about Summer 1944 and on were

The rear fuselage of the DFS *228 V1* with the *HWK 509D* combustion chamber exposed. Notice that the vertical stabilizer is gone...broken off or merely removed. Notice, too, that the combustion chamber has support rods leading to the bulkhead from the chamber's front.

never flight tested at Ainring to check their ability to maintain pressure. Too many Allied fighters hanging around. So, our difficulties had to do with keeping the cabin pressurized. No, *Mödel* did not design the cockpits, but he did construct them. We had some thirty engineers at *DFS*-Ainring working on these *DFS 228/346* projects between 1944-1945.

MYHRA - So the *DFS 228* was flown only as a sailplane? Did the *DFS 228* ever use another type of cockpit cabin?

NIESS - I do not remember what type of cockpit arrangement the *DFS 228* was flown with, but the wooden cockpit model which you see in these photographs, well, it was never flight tested at all on the *DFS 228* or any other aircraft prior to the surrender. Like I said, we had many, many difficulties with it, and they all stemmed from the fact that we had to build it out of wood. We could not make the cabin tight enough to hold the required pressure needed to fly the machine at high altitude. The first *DFS 228* was actually constructed in 1941 and mostly out of wood. It was to be used at heights up to 10 to 12 kilometers [32,000 to 40,000 feet]. Prior to flying the *DFS 228*, *Erick Klöckner*, a *DFS*-Ainring staff test pilot, made a test flight in another sailplane with a pressure cabin up to 11 kilometers [36,000 feet] in 1940. This was a record-setting flight at that time, I believe. By the way, I have been noticing the dates on your drawings of the *DFS 228*. They are all mostly dated November 1944. Some of the drawings show dates of August 1944, and some have my initials on them. I was of the opinion that all these documents and drawings were destroyed in the war.

MYHRA - No, not all documents from this period. Some photographs and drawings exist today in aviation museums and private collections. From time to time I am fortunate to find them. People holding private collections have been very good to me, because they have allowed me to make copies with my camera. You worked only on the construction of *DFS 228* and *Me 328?*

NIESS - Yes. *Felix Kracht* was my boss, and he was the chief of *DFS*-Ainring's research department. *Meyer* was the *DFS*-Ainring's leader for aerodynamics.

End of *Hans Niess* Interview

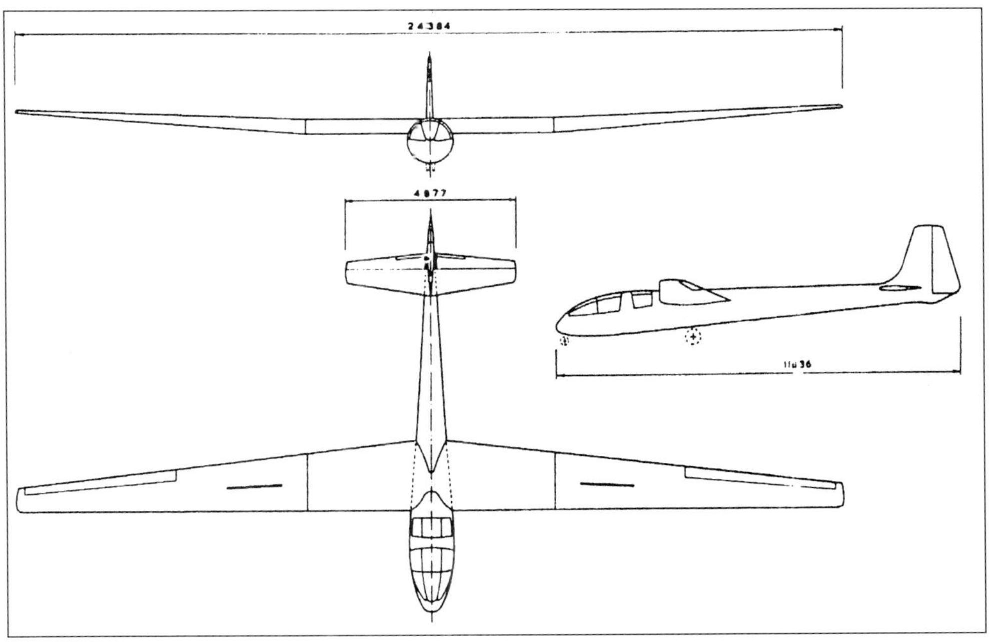

A pen and ink drawing of the proposed *Slingsby T-44,* a high altitude reconnaissance sailplane with a pressurized cockpit cabin. *Slingsby Sailplanes, Ltd.* had studied the *DFS 228* while it was at RAE-Farnborough, England. The machine may have in fact been given to *Slingsby* by the RAE-Farnborough. Nevertheless, about 1947, the *DFS 228* was scrapped...either by RAE-Farnborough or by *Slingsby.* Three view drawing courtesy: *Norman Ellison, British Gliders and Sailplanes...1922-1970.*

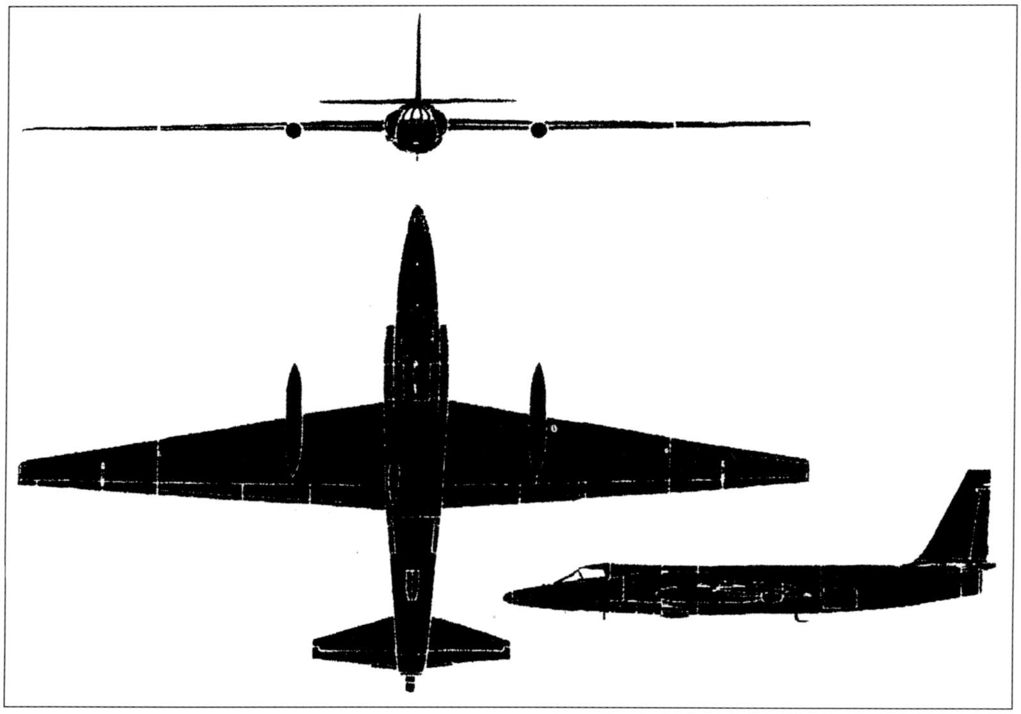

A three view reverse negative image of the *Lockheed U-2*. Note the similarities in style but then again each machine had pretty much the same mission: ultra high altitude photo reconnaissance.

An artist's rendering of a *Lockheed U-2* coming under fire from "*SAM*" missiles high over the USSR. The *DFS 228* and the *U-2* were both designed to do the same type of job...ultra high altitude photo reconnaissance. If the *DFS 228* looks like a rocket powered sailplane then the *U-2* looks even more so with its high aspect ratio wing.

A wind tunnel scale model of *Felix Kracht* and *Eberhard Meyer's* speed of sound *HWK 509D* bi-fuel liquid rocket powered *DFS 346* machine. This photo comes form the USSR postwar when the Soviets placed the model in one of their wind tunnels to determine the machine's aerodynamics and how well it would handle forces up to and beyond the speed of sound.

The long prob extending out of the *DFS 346 V1's* cockpit canopy gives it purpose away...that is, to measure and record data during high speed attempts. With its swept back wings and tail plane, the *DFS 346 V1* was designed to reach and travel beyond the speed of sound or *Mach 1.0*. It wasn't flight ready at war's end but the Soviets took the machine and scores of German aircraft designers and engineers formerly with *Junkers, Siebel,* and *Heinkel* to a location outside of Moscow. In its only high speed record run, the machine failed to break the speed of sound. Instead, it went out of control and its pilot *Wolfgang Zeise*, had to leave it. However, all the safety devices built into the machine by *Felix Kracht* worked as planned and *Zeise* parachuted back to earth safely. Digital image by *Mario Merino*.

Opposite: The speed of sound machine known as the *DFS 346 V1*. Designed by *Felix Kracht* and his able assistant *Eberhard Meyer*, this aircraft was powered by a single *HWK 509D* bi-fuel liquid rocket engine with two combustion chambers of equal size. The *346 V1* is seen from its nose starboard side. Digital image by *Mario Merino*.

A direct port side view of the *DFS 346 V1*. Notice that the pilot has both combustion chambers operating. The aircraft is unpainted and its various aluminum panels providing different shades and hues. Digital image by *Mario Merino*.

The *DFS 346 V1* as seen from beneath. Notice that only one of the two combustion chambers (upper) is providing thrust at this moment. Digital image by *Mairo Merino*.

Index